FLORIDA CITRUS DIAGNOSTIC GUIDE

Harold W. Browning
Robert J. McGovern
Larry K. Jackson
David V. Calvert
Wilfred F. Wardowski

Florida Science Source, Inc.
Lake Alfred, Florida

Library of Congress Catalog Card Number: 95-60087

ISBN 0-944961-01-0

FLORIDA CITRUS DIAGNOSTIC GUIDE

Current printing (last digit): 10 9 8 7 6 5 4 3 2 1
Published in the United States of America

CONTENTS

AUTHORS

Harold W. Browning, Associate Professor, Citrus Research and Education Center, 700 Experiment Station Road, Lake Alfred, Florida 33850

Robert J. McGovern, Associate Professor, Southwest Florida Research and Education Center, P.O. Drawer 5127, Immokalee FL 33934-5127

Larry K. Jackson, Professor Emeritus, Citrus Research and Education Center, 700 Experiment Station Road, Lake Alfred, Florida 33850

David V. Calvert, Professor, Agricultural Research and Education Center, P.O. Box 348, Fort Pierce, Florida 34954-0248

Wilfred F. Wardowski, Professor, Citrus Research and Education Center, 700 Experiment Station Road, Lake Alfred, Florida 33850

The authors are all members of the University of Florida, Institute of Food and Agricultural Sciences.

Florida Science Source, Inc.
Lake Alfred, Florida

PREFACE

This text provides color photographs for the identification of Florida citrus insects, mites, diseases, physiological disorders, horticultural problems and nutritional deficiencies and toxicities. It has been many years since a comprehensive book has been available which attempts to cover these subjects in this manner. The text by Dr. R. M. Pratt listed as a reference was published 38 years ago, has been out-of-print for many years, but is still in wide use today. Dr. Pratt's book was an inspiration for this effort, although the text and photographs are all new. We hope that this book will help fill the need for a handy Florida citrus reference.

Scientific names are used where possible to identify organisms. In addition, common names used in Florida and more generally are provided where available. However, some species do not have individual common names. It is also recognized that common names vary by region and thus may not be definitive without reference to the scientific names.

Illustrations used in this guide attempt to portray the general appearance of the organisms and their injury on citrus leaves, fruit, etc. Variation in the appearance, coloration and other features of individual pests, and also the damage caused makes diagnosis difficult at times. In some cases, submission of specimens or samples of damage to experts may be necessary to characterize the problem. Pests and disorders commonly occur together, and thus diagnosis of more than one organism or disorder may be necessary. Additional resources should be consulted to provide more comprehensive information on individual pests, diseases and disorders and their management. Some of these resources have been listed at the end of the book.

Florida Citrus Diagnostic Guide is not designed to recommend pesticide or fertilizer uses, although general comments will be found. Suggestions are made for cultural practices which may increase or decrease the severity of the pest, nutrient imbalance or other problems. Consult county Extension offices for current nutrient and pest management recommendations.

The authors are indebted to the generous permission of many people who allowed their photographs to be included in this book. These people are noted in the respective figure captions. Special recognition is noted for Robert L. Phillips, Okeechobee, Florida, who devoted many hours and talent for several photographs including the cover illustration. Terry Zito provided photographic support to the authors, and provided a valuable service with advice and skill.

Numerous reviewers improved and refined the manuscript as this work progressed. The authors are responsible for any errors in this book and would appreciate your comments to improve future printings.

Harold W. Browning
Robert J. McGovern
Larry K. Jackson
David V. Calvert
Wilfred F. Wardowski

IDENTIFYING INSECTS AND MITES

Paramount to the development and implementation of management for pests affecting citrus or other agricultural production systems is the identification of the causal agent through injury assessment, sampling, or often just simple field observation. Diagnosis of the causative agents will lead to understanding of the seasonality of injury, the population densities of importance, and the appropriate approaches to management where necessary. Thus, accurate diagnosis and characterization of the specific pest is a fundamental entry point for pest management.

The complex of arthropod pests affecting citrus in Florida is diverse, with species of mites and insects representing a wide range of habits, life histories, and resulting forms of damage. Species are adapted to exploit roots, structural components of trunk and limbs, foliage, and fruit. Some are direct feeders on fruit, causing cosmetic injury to fruit surfaces or leaving evidence of their presence on fruit at harvest. Other species feed on leaf tissue with chewing mouthparts, through piercing leaf cells and extracting cell contents, or through probing into xylem or phloem tissue to extract plant fluids. Size ranges from large insects such as katydids and grasshoppers to minute insects and mites that require a hand lens to observe.

Groups most commonly encountered can be divided according to their feeding sites, their injury, the taxa to which they belong, or the relative importance that they have to citrus production or more precisely, to integrated pest management. New arthropod pests are finding their way to Florida citrus at an alarming rate, so the complex of pests is ever expanding.

Furthermore, many of the arthropods observed in Florida citrus should not be classified as pests, but merely as components of a perennial citrus cropping system. For some of these species, only rarely do they cause sufficient damage over a large enough area to constitute pest status, and it is important to distinguish consistently injurious species from those who merely coexist with the other elements of citrus production.

Among the chronic pests are several mite species (Fig. 1-7). Mites of agricultural importance in Florida citrus possess a wide range of characteristics. Whereas, most mite groups conform to the normal case of 4 pairs of legs, similar to spiders, certain groups such as the rust mites possess only two pairs of legs. Feeding strategies range from foliage to fruit to stems to combinations of one or more substrata, and some are known to move from one substrate to another as the season progresses. Still others are beneficial in that they are predaceous on soft bodied insects and mites, and thus serve as natural control agents (Fig. 57).

Another group of arthropods well represented in Florida is the order homoptera. More than 50 species of homopterans feed and reproduce on citrus, and many of these have been highlighted in this diagnostic guide (Fig. 8-35). Biological diversity and specialization among groups of homopterans make them difficult to understand at a glance. Production of honeydew, reproduction without males, sessile life habits, and other interesting features complicate characterization of species. To assist in understanding the homoptera, the opposing table is provided to summarize fundamental differences among the groups of homoptera most commonly encountered. Reference to this table will provide an entry point for identifying the family to which species belong, after which closer examination may allow species identification.

Many other groups are included, and characteristics useful in their identification are provided in the species descriptions. Among these are the hemipterans (Fig. 35-37), closely related to the homoptera. These "true bugs" are most often encountered feeding on mature fruit. The order Lepidoptera is represented by two species (Fig. 38-39) that are foliage feeders. Root weevils are well represented by the species attacking Florida citrus (Fig. 40-44). These curculionids have similar habits in that adults feed on young leaves, oviposit on leaves or under fruit calyces, and larval and pupal development occurs in soil, where larval feeding on roots causes the primary damage. Orthopterans include katydids, grasshoppers, and a new species to Florida, the Asian cockroach (Fig. 45-48).

While this diagnostic guide does not cover all species of arthropods likely to be found in Florida citrus, it does provide information and diagnosis on the majority of species that consistently are encountered. Additional species of thrips, coffee bean weevils, and others are recognized as present and at times of potential economic consequence to citrus, but these species are not included here.

CHARACTERISTICS TO SEPARATE HOMOPTEROUS INSECT PESTS

GROUP	IMMATURE FORM	IMMATURE COVER	ADULT FORM
APHIDS	Immatures fully developed, with legs, eyes, antennae.	No protective covering, naked.	Both sexes with winged adults, not present in all generations.
WHITEFLIES	Immature hatches as active crawler, then settles to develop as sessile form; legs, eyes, and antennae reduced. Honeydew produced	Generally no protective covering, some immatures with waxy filamentous covering.	Both sexes with winged adults.
ARMORED SCALES	Immature hatches as active crawler, then settles as soft sessile form; legs, eyes, and antennae reduced. No honeydew production.	Membranous immature covered with hardened "shell" secreted by insect; can be lifted off to expose soft immature.	Adult males, when present, are winged; females retain immature form beneath armor covering.
SOFT SCALES	Immature hatches as active crawler, then settles to feed. At molts and when disturbed, immature can relocate. Honeydew produced.	Body of scale may be somewhat hardened or covered with wax; no separate "shell" as with armored scales.	Adult males, when present, are winged; females retain immature form and oviposit beneath body.
MEALYBUGS	Immature hatches as crawler and retains mobility; retains legs, eyes and antennae.	Body covered with wax filaments.	Adult males winged; females retain immature form and are mobile; eggs are in wax covered sacs.

CITRUS RUST MITE, *Phyllocoptruta oleivora* (Ashmead)

Citrus rust mite (Acari: Eriophyidae) is the most important arthropod citrus pest worldwide. Individual mites are extremely small. Adult mites are slightly over 0.1 mm in length, and are about 3 times longer than wide, tapering from front to rear (Fig. 1). They have only 2 pairs of short legs. Body color is light yellow, and the 2 immature stages closely resemble adults. Clear non-textured eggs are deposited singly or in small clusters on leaf and fruit surfaces.

Citrus rust mites infest leaves and green twigs of all citrus varieties, but are most important in feeding on developing fruit. Feeding is accomplished through piercing plant epidermal cells by the mouthpart stylets. Visual evidence of feeding on fruit rind cells is discoloration or russetting. The extent and appearance of russetting varies with the time of infestation, with early fruit infestation resulting in surface bronzing. Rust mite infestation later in fruit development results in dark brown to black scarred areas (Fig. 1). A light dusty fruit appearance often accompanies heavy feeding, which represents buildup of tremendous numbers of cast exoskeletons as the mites molt through 2 immature stages to the adult form. In addition to reduction in cosmetic appearance caused by feeding, sustained heavy populations can reduce fruit size and thus reduce yields. Populations occur all year on citrus trees, with populations utilizing different parts of the tree at varying times of the year. Fruit are not uniformly infested, as rust mites prefer fruit exposed to sunlight but avoid the hot spots on outer canopy fruit. Massive populations can develop quickly. One female rust mite lays 1 or 2 eggs per day over a period of about 20 days, contributing up to 20 to 30 new rust mites to the population, and rust mites can complete a generation in about 10 days. Site history of rust mite infestation provides some indication of population potential as a new season begins.

Citrus rust mite is attacked by a complex of natural enemies, but generally under Florida conditions, beneficial organisms are unable to consistently maintain rust mites below damaging levels. There is, however, an effective fungus, *Hirsutella thompsonii* Fisher, which can respond under appropriate environmental conditions to infect and dramatically reduce high rust mite populations.

Figure 1. Valencia orange fruit damage due to infestation by citrus rust mite, *Phyllocoptruta oleivora* (top); and citrus rust mite adults on leaf surface (bottom).

PINK CITRUS RUST MITE, *Aculops pelekassi* (Keifer)

The pink citrus rust mite (Acari: Eriophyidae), is a lesser known but relatively common mite pest on citrus in Florida. Similar in appearance and biology to the well known citrus rust mite, *Phyllocoptruta oleivora* (Ashmead), the pink citrus rust mite can be found throughout the state. In addition to fruit russetting and leaf stippling typical of citrus rust mite, pink citrus rust mite is capable of inducing other leaf symptoms, including leaf distortion, mesophyll collapse, and leaf drop.

Adult pink citrus rust mites are predominantly females, with males only represented in about 10% of the population. The mites are extremely small, but under hand lens magnification of 10X or higher, the mites appear elongated, broader anteriorally with sides tapering towards the posterior (Fig. 2). Coloration is pink, whereas, citrus rust mite is yellowish-beige in color. Under higher magnification, the dorsal anterior shield of pink citrus rust mite can be seen as more strongly developed with clearly defined plates. Eggs of pink citrus rust mite are translucent white in contrast to opaque eggs in citrus rust mite. Pink citrus rust mite eggs are typically scattered across leaf and fruit surfaces, while citrus rust mite eggs are more localized along leaf midribs or fruit depressions. This does not reflect adult mite distribution, since citrus rust mite can be found scattered on the leaf. Pink citrus rust mite adults are often observed along leaf margins, making aerial dispersal more likely than for the citrus rust mite. This phenomenon is further supported by the habit of pink citrus mite adults of rising upright onto pads located at the posterior of their body which also promotes lifting the minute mite into air currents.

Pink citrus mite reproductive biology is similar to that of citrus rust mite, with short generation times under favorable climatic conditions, particularly those found in greenhouse citrus production. The propensity to develop locally high densities on fruit and foliage is shared with citrus rust mite. This species is more attracted to young leaves rather than the mature leaves favored by citrus rust mite and also appear to prosper on loose-skinned fruit varieties.

Figure 2. Pink citrus rust mite adults (top, Courtesy C. W. McCoy); and severe infestation on leaf (bottom, Courtesy C. W. McCoy).

CITRUS RED MITE, *Panonychus citri* (McGregor)

Citrus red mite (Acari: Tetranychidae), or red spider, as it is sometimes referred, is a widely distributed spider mite in Florida citrus. Females of this species are easily distinguished from other spider mites on citrus by their rounded body (0.3 to 0.5 mm), their deep red to purple coloration, and the presence of white bristles (setae) which arise from reddish tubercles on the dorsal body surface (Fig. 3). Male citrus red mites are long legged, and are much smaller, flatter than their counterparts. The reddish eggs of citrus red mite are round and possess a central dorsal stalk from which several fine silken fibrils radiate and attach to the substrate (Fig. 3). Eggs hatch into small 6-legged red larvae, and following the first molt, there are 2, 8-legged nymphal stages. The immobile interval between active stages, during which the molt occurs, is more or less equal to the active stages. An opaque cast also characterizes this stage.

Citrus red mite is primarily a leaf inhabiter, preferring upper over lower leaf surfaces. Varietal differences in red mite susceptibility have been noted, with lemons and grapefruit supporting larger populations than oranges. Other host plants include rose, pear, and woody ornamentals. Feeding by citrus red mite is typical of spider mites. Visual evidence of feeding is seen as stippling and some silvering of upper leaf surfaces. Sustained heavy feeding by citrus red mite can lead to more serious leaf symptoms, including the loss of leaf moisture known as mesophyll collapse and the sudden desiccation and abscission of leaves from otherwise healthy stems, known as “firing” (Fig. 3). These symptoms are often induced or exaggerated by dry, windy conditions. Stippling may also be visible on fruit which have sustained heavy mite feeding.

Populations of citrus red mite are rather sporadic in Florida and have been observed to build at almost any time of year when temperature and humidity conditions favor survival and reproduction. Life cycle times are as short as 11 to 12 days and females lay up to 40 eggs over a lifetime. Natural enemies have been associated with citrus red mite, but as with other mite species in Florida citrus, red mite populations often explode despite the presence of these natural enemies.

Figure 3. Adult female of citrus red mite, *Panonychus citri* (top left), on citrus leaf; eggs of citrus red mite along leaf midrib (top right); and firing of Valencia foliage due to cumulative citrus red mite feeding (bottom).

SIX-SPOTTED MITE, *Eotetranychus sexmaculatus* (Riley)

Another species of spider mite found on citrus, the six-spotted mite (Acari: Tetranychidae), is distinctive in appearance and biology. Females of this spider mite are rounded, yellow in color and display 1 to 3 pairs of dark spots located along the dorsal surface of the body (Fig. 4). Long bristles (setae) also are visible emerging from the dorsal surface of the mite. Six-spotted spider mite is somewhat smaller than other species on Florida citrus. Males are smaller than females and are teardrop shaped, with legs longer in proportion to the body than those of females. Immature mites are similar in color and appearance to adults, and undergo 3 molts typical of spider mites. As with most mites, the larval stage emerging from the egg has 3 pairs of legs, while subsequent nymphal and adult stages possess 4 pairs of legs. Eggs are pale yellow and are similar in form to those of citrus red mite, bearing a dorsal stalk attached to thin silk strands which span from the egg to the substrate.

Six-spotted spider mite shows characteristic behavior in utilizing undersides of leaves almost exclusively, strongly orienting to the midrib area. Extensive webbing is produced by this species, and often eggs are attached to the webbing. Discoloration of the feeding site occurs in response to cell damage, and the leaf surface collapses inward, causing a bulging of the leaf toward the upper surface. The yellow shiny coloration contrasts with healthy green adjacent tissue. Sustained feeding results in sufficient damage to induce defoliation, and severe leaf distortion is often observed when young leaves are infested.

Populations of six-spotted mites occur on citrus year around, but populations vary in their location on trees through the seasons. Winter populations generally are found on interior foliage, while spring infestations can be found on new flush. Although found throughout the state on citrus, six-spotted mite is not consistently a problem. Highest populations generally are found in late winter and spring. Fruit infestation is not common, but under high pressures, mites may feed on fruit surfaces, causing some discoloration and premature fruit drop.

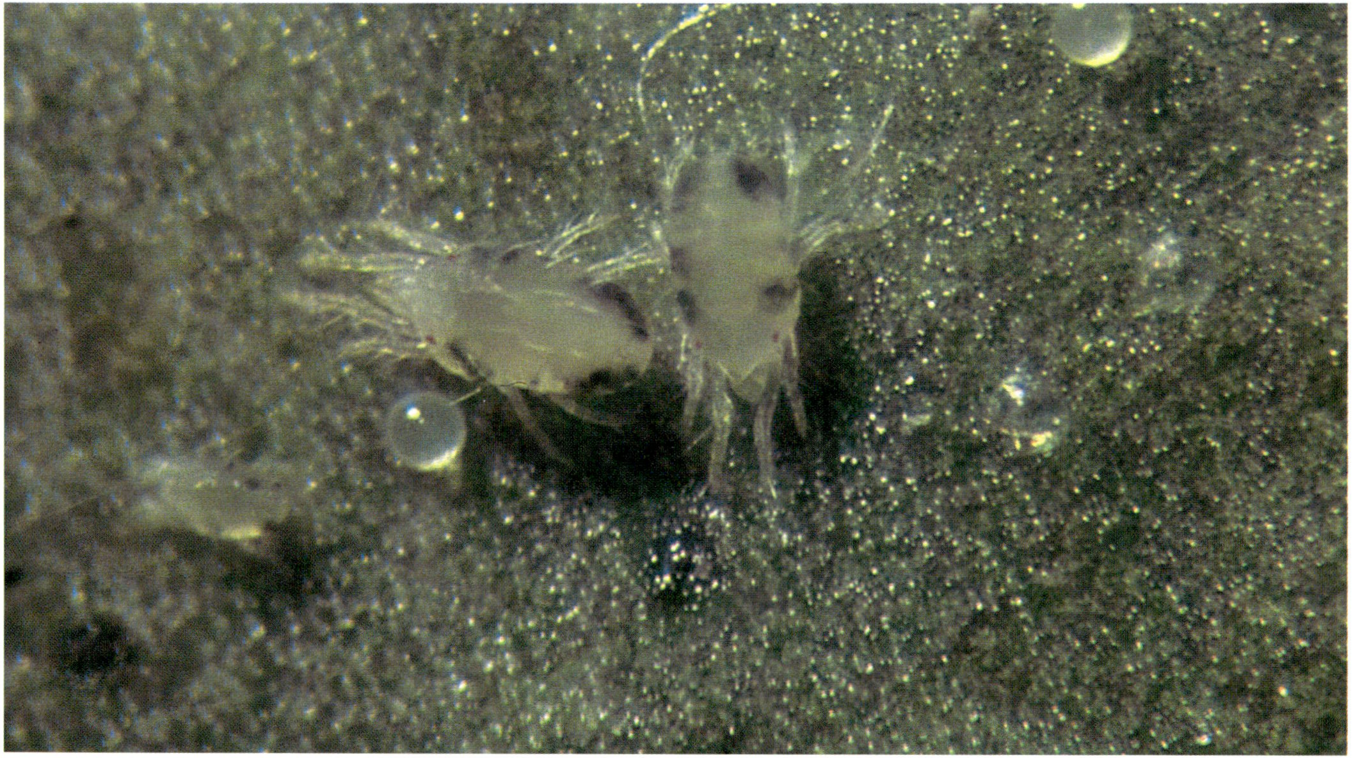

Figure 4. Six-spotted spider mite eggs, nymph and adult females on citrus fruit (Courtesy M. M. Abou-Setta).

TEXAS CITRUS MITE, *Eutetranychus banksi* (McGregor)

Texas citrus mite (Acari: Tetranychidae), is the third species of spider mite and probably the most recent of the 3 to impact on Florida citrus. This species shares many features with citrus red mite, *Panonychus citri* (McGregor), but upon examination can be distinguished in several ways. Adult females of Texas citrus mite are yellowish-green in color, broadly oval and similar in size to citrus red mite females (Fig. 5). Some internal dark spots may be seen from the dorsal body surface, but these spots are quite distinct from the pigmentation spots on the back of six-spotted mites, *Eotetranychus sexmaculatus* (Riley). Short, coarse bristles (setae) are present on female Texas citrus mites in contrast to longer white bristles on citrus red mite (Fig. 3) or the long thin bristles on six-spotted mite (Fig. 4). Male Texas citrus mites are similar in color, but smaller, flatter, and tearshaped, and have exceptionally long legs for spider mites (Fig. 5). Larvae and nymphs of Texas citrus mite resemble the adults in color and shape (Fig. 5). Eggs are discshaped, and change from yellow to tan or green as hatching approaches.

Texas citrus mites prefer upper leaf surfaces for infestation. Eggs are laid along leaf midribs and veins, and developing mites tend to feed on upper surfaces. Little or no webbing is associated with colonies of Texas citrus mite. Feeding injury occurs as with other spider mites via cell penetration and cell content extraction. The symptoms of Texas citrus mite are very similar to citrus red mite, with stippling, mesophyll collapse (Fig. 5) and firing often observed. Mites also feed on fruit but generally without inducing direct visible damage to the fruit rind.

Texas citrus mites occur throughout the year in Florida citrus, but appear to prosper under low humidity conditions. Populations build in spring and fall more than during rainy summer months. Generation times are similar to other spider mite species, allowing rapid population buildup under favorable conditions. Natural enemies are associated with Texas citrus mite, but like other mite pests on citrus, the Texas citrus mite is able to overcome natural enemy pressure and build to high populations which require pesticidal suppression in some situations.

Figure 5. Nymphs at top, and transitional adult female Texas citrus mite on leaf surface (top left); male Texas citrus mite (top right); and mesophyll collapse of leaf tissue caused by Texas citrus mite feeding (bottom).

BROAD MITE, *Polyphagotarsonemus latus* (Banks)

The broad mite (Acari: Tarsonemidae), is a common tropical or subtropical mite with an extensive host range which includes numerous agricultural and ornamental crops. Broad mite appearance varies considerably by age and sex. Adult female mites are 1 to 3 mm in length, broadly oval, highly convex, with color varying from white to dark green. Adult mites appear hard and shiny. Members of this family are characterized by specialized leg structures and some of these structures are apparent in broad mites. Young females are more subcircular and lighter in color than mature females, but are otherwise similar in appearance. Males are shorter, tapering to the posterior, and have 4 pairs of legs which are visibly longer than those of the female. Males hind legs are modified into claw-like appendages, and body coloration progresses from transparent to amber as the mite matures.

Broad mite eggs are elongate-oval and are tightly attached singly to the plant substrate in leaf and fruit depressions. These large eggs have tops sculpted with rows of nodules (tubercles) which appear white in contrast with the clear cast to the remainder of the egg. Eggs hatch into active 6-legged larvae, after which an inactive (sessile) stage is passed. Adults emerge following splitting of the pupal skin.

Broad mites feed on citrus leaves and fruit, being primarily found on the undersides of leaves. Leaf damage becomes visible as young leaves expand, and is manifested as scorching, distortion and discoloration. Effects tend to appear on tender young growth, stimulating buds near the damage to break apical dominance and produce multiple bud growth. Feeding on citrus fruit, particularly lemons, results in discoloration (Fig. 6).

Populations of broad mite tend to be localized, as adult female mites tend to remain on a single leaf as they mature. Generation times for this species are extremely short, with populations reproducing in about 5 days under favorable conditions. Reproduction occurs year around, with developmental times expanding as temperatures decrease. Environmental conditions which appear to favor broad mite development include warm temperatures, high humidities, and filtered light.

Figure 6. Adults, nymphs and eggs of broad mite on citrus leaf (top left); young leaf distortion due to high populations (top right); and immature lime fruit with extensive broad mite injury (bottom).

FLAT MITES, *Brevipalpus* spp.

These slow moving, flat mites (Acari: Tenuipalpidae), are related to spider mites and thus somewhat resemble spider mites (Family Tetranychidae). Tenuipalpidae are often referred to as false spider mites. Species of *Brevipalpus* can cause serious injury to a wide range of agricultural crops, and have been reported on Florida citrus for several decades. Three species which are known to attack citrus in Florida include *B. californicus* (Banks), *B. obovatus* Donnadieu, known as the privet mite, and *B. phoenicis* (Geijskes).

All flat mite species on Florida citrus are red or orange, very flat, slightly smaller than spider mites and are very slow moving, making them difficult to detect at low populations (Fig. 7). They generally feed on undersides of leaves, but may also be found on developing fruit and on bark. Adult females are teardrop shaped, very flat, and possess 4 pairs of stout wrinkly legs. The dorsal body surface is reticulated, although this feature may not be apparent under field observation. Short bristles (setae) are present along the posterior margins and central portion of the body. Adult males are similar in appearance but far less common. Eggs are elliptical and orange to red, according to species, and generally adhere to plant surfaces by a sticky adhesive. Eggs are laid at a rate of one per day and hatch into reddish 6-legged larvae after more than 1 week. Immature development progresses through larval and 2 nymphal stages with quiescent stages between each. Total developmental time ranges from 10 to 26 days at moderate to warm temperatures, and coupled with low egg laying rates, results in slow population increase. Under Florida conditions, flat mites normally do not develop populations which inflict economic damage to citrus.

Feeding by these mites varies widely by species, plant part and location in the world. Leaf feeding may result in tiny gummy deposits, diffuse chlorotic spotting, necrotic ring spots (privet mite) or in extreme cases, may result in the injection of toxins which result in nailhead rust or leprosis, a scaly, pitted appearance. Lesions on bark resulting from *Brevipalpus* feeding may expand as the twigs expand, causing twig dieback. Similarly, fruit damage ranges from light stippling to rind spotting to severe lesions and in certain citrus producing countries, extensive fruit drop. *B. californicus* and *B. phoenicis* have been implicated in these more severe symptoms.

Figure 7. Flat mite colony on immature orange fruit with clusters of eggs and adult females (Courtesy M. M. Abou-Setta).

FLORIDA RED SCALE, *Chrysomphalus aonidum* (L.)

Florida red scale, *Chrysomphalus aonidum* (Homoptera: Diaspididae), is an armored scale which was a severe pest until 1959 when the parasitic wasp *Aphytis holoxanthus* DeBach was introduced.

Adult female Florida red scale covers are roughly circular in shape (about 2 mm diameter), slightly convex and dark brown or black as viewed from the top (Fig. 8). Beneath the scale cover is the soft, bright yellow body of the female scale (Fig. 8). Immature males are found beneath armor covers less than half the size of female covers, and are slightly oval and with a tan-brown fringe (Fig. 8). Males emerge from the scale covers as winged gnat-like adults. Early nymphal armor is incorporated into subsequent armor at molts and appears as concentric rings.

Mated reproductive females produce bright yellow eggs which are deposited in groups of up to 10 or so at a time under the scale armor (Fig. 8). These eggs generally hatch within 24 to 48 hours. Females lay an average of 145 eggs over their lifespan, and development from egg to adult can occur in about 40 days under favorable conditions. Six or more generations per year occur on citrus in Florida.

Florida red scale has a tremendous host range, attacking more than 600 host plants. Feeding on citrus is restricted to leaves and fruit. Populations appear to explode on developing fruit from very low leaf populations, and at harvest numerous adult scale renders the fruit incapable of meeting fresh fruit marketing standards. Scale feeding on the fruit results in discoloration at feeding sites. Under sustained high populations, defoliation and twig dieback occurs.

Populations of Florida red scale generally are kept below damaging levels by the combined actions of *Aphytis holoxanthus* and other natural enemies. However, sporadic population increases are occasionally observed, often in concert with repeated application of broad spectrum pesticides, which are presumed to interfere with natural enemies.

Figure 8. Florida red scale females, males, and immatures on citrus fruit (top); and adult female Florida red scale with armor removed to show soft body, eggs, and crawlers (bottom).

PURPLE SCALE, *Lepidosaphes beckii* (Newman)

Purple scale (Homoptera: Diaspididae), has declined from being one of the most important citrus pests in Florida to minor importance in the last 3 decades through successful biological control. Today, purple scale is widespread at extremely low densities in Florida citrus. The adult female scale is 2 to 3 mm long, oyster shell shaped and dark brown or purple (Fig. 9). The visible portion of the insect is the chitinous armor, and the scale can be seen to be pearly white beneath the scale cover, elongate and tapering outward towards the posterior. A thin sheath of cover material partially or completely envelopes the underside, separating the scale from the plant surface upon which it is feeding.

Male scale are similar in shape but less than 1/3 the size of the mature female (Fig. 9). Male scale armor is the same color as the female's, and beneath the metamorphosis to a winged, legged adult takes place. Adult males do not feed and live only 1 or 2 days. Males are assisted in locating receptive virgin female scale by the release of pheromones by the females.

Mated purple scale females oviposit a clutch of 40 to 80 pearly white elongate eggs arranged in 2 rows beneath the scale cover. The membranous female body recedes within the scale cover as the eggs are laid. Hatching white crawlers exit the armor of their mothers and settle on suitable substrata and begin feeding. The scale is subsequently incapable of mobility. This generalized cycle is typical of all armored scale insects.

Purple scale infests leaves, fruit, green twigs, older twigs, and the bark of branches and the trunk. On leaves, purple scale feed on cell contents, resulting in yellow discoloration which spreads as the population continues. Under heavy population pressure, leaves drop and twigs suffer dieback. Fruit infestations can be heavy (Fig. 9), and as infested areas resist color transition from green to yellow or orange at maturity. Generation times are moderate, with purple scale requiring 40 days or more to complete development from egg to adult. Several overlapping generations occur each year under Florida conditions.

The introduction of the parasitic wasp *Aphytis lepidosaphes* Compere in the late 1950's has brought about complete biological control of purple scale.

Figure 9. Group of purple scale on fruit surface composed of females and males (top); and mature grapefruit heavily infested with purple scale (bottom).

GLOVER SCALE, *Lepidosaphes gloveri* (Packard)

Glover scale (Homoptera: Diaspididae), also known as long scale, is often associated in mixed population on citrus with purple scale, *Lepidosaphes beckii* (Newman), to which it is closely related. Slightly more brown in color than purple scale, the armor shape of glover scale is generally distinctive from that of purple scale. Whereas, purple scale is oyster shell shaped, glover scale is elongate, relatively straight sided, and tapers slightly toward the anterior end (Fig. 10). The exuviae of the first nymphal instar is located at the anterior end of the scale and is yellowish in color. Glover scale is narrower than purple scale and approximately 3 mm in length. Beneath the armor, the scale body is pearly white. White elongate eggs are deposited in 2 rows within the armor cover as is the case with purple scale.

Glover scale infests leaves, fruit, twigs and the woody bark of branches. The scales seem to be more or less randomly oriented on leaf and fruit surfaces, and only occasionally are clustered. On woody bark, the scale are often oriented with the grain of the bark and thus are parallel with one another (Fig. 10). Due to the brownish color of scale armor, glover scale is difficult to detect on bark and often goes unnoticed.

Populations of glover scale rarely build to damaging levels in Florida citrus, although individuals occasionally appear on fruit harvested for fresh market sales. Twig dieback has been observed in association with high populations of glover scale. Parasitic wasps in the genus *Aphytis* are associated with the scale and may be in part responsible for the minor pest status of this scale in Florida.

Figure 10. Adult female glover scale on citrus leaf (top); and female glover scales aligned with citrus stem (bottom).

CITRUS SNOW SCALE, *Unaspis citri* (Comstock)

Citrus snow scale (Homoptera: Diaspididae), is one of the bark inhabiting armored scale insects, and the white snowy appearance of high populations of male scale on citrus tree trunks and branches give the scale its name. This scale is spread throughout the state where it is a sporadic pest which goes unnoticed until populations reach extremely high levels.

Adult females of citrus snow scale are very similar in appearance to female purple scale, *Lepidosaphes beckii* (Newman), being dark brown to purple in color and shaped like an oyster shell (Fig. 11). However, a longitudinal ridge runs the length of the snow scale female cover. Beneath the female snow scale armor is an elongated orange-yellow membranous scale body. From 1 to 4 yellowish eggs are laid, and egg duration is limited to a few hours. Yellow crawlers leave the female cover to settle and begin feeding. Immature male citrus snow scale covers are very different from female covers, being about 1/3 the size and white in color. The male scale cover is composed of a terminal remnant of early nymphal exoskeleton from which a 3-part waxy cover for the developing pupa protrudes (Fig. 11). Following pupation, the winged orange adult scale will emerge to begin seeking mates. Discreet holes visible in the cover of male citrus snow scale are evidence of parasitization of the scale by the wasp, *Encarsia.*

Populations of snow scale are most often observed on tree trunks, with individual scales forming dense masses virtually on top of one another (Fig. 11). Over time, the trunk may take on a dull white appearance from the disproportionately male-biased population of scales present. Females are nearly invisible on bark due to coloration. About 60 days are required for completion of a generation of citrus snow scale.

Sustained heavy feeding by citrus snow scale appears to be associated with splitting of trunk and branch bark, an effect of repeated cell damage to active bark epidermis. Fruit and leaf infestations are incidental although it has been shown that snow scale can successfully complete development on leaves and fruit of orange and grapefruit. The citrus snow scale does not appear to be regulated by natural enemies and occasionally reaches populations of concern to producers.

Figure 11. Citrus snow scale female and several males on citrus leaf (top); and lime tree branch heavily encrusted with citrus snow scale (bottom).

LESSER SNOW SCALE, *Pinnaspis strachani* (Cooley) and FERN SCALE, *Pinnaspis aspidistrae* (Signoret)

Lesser snow scale (Homoptera: Diaspididae), and the closely related fern scale are often confused with citrus snow scale, *Unaspis citri* (Comstock) on citrus, but there are characteristics which readily separate these species. The similarity between the species is based on the 3-ridged white waxy covering which protects the developing male scale in all 3 cases. The size of the immature male is similar, and based on male morphology alone, it is difficult to make a species determination. However, the female scale are distinctive and thus the species are discernible. Whereas, the female scale cover of lesser snow scale is oyster shell shaped, the color is white or dirty white and without longitudinal ridges (Fig. 12). Females of fern scale are elongate, thick, and are brown in color (Fig. 12). Citrus snow scale females are similar in shape, but the scale cover is dark brown to purple and has a longitudinal ridge across the dorsum of the scale (Fig. 11). Immature scale are yellowish brown and more circular than adult females.

The other distinction between lesser snow scale, fern scale and citrus snow scale is the locational preference exhibited by each species. Lesser snow scale attacks leaves, fruit, and bark of citrus, but is most commonly encountered on leaves and fruit. Fern scale also favors leaves and fruit for infestation. In contrast, citrus snow scale is most often encountered on mature woody bark of trunks and scaffold branches. Furthermore, localized colonies of lesser snow scale and fern scale generally are composed of about 30 males and 1 or 2 females, oriented on the leaf or fruit to form a white patch. Citrus snow scale tends to be quite scattered at low densities and without a set pattern on bark at high densities.

Lesser snow scale and fern scale are distributed throughout Florida but are more common in southern regions of the state. In addition to citrus, palms, *Hibiscus* spp., ferns, and other ornamentals are infested by *Pinnaspis* species. Populations rarely reach damaging levels on citrus, and the presence of patches of males and a female do not discolor fruit. Standard packinghouse washing operations tend to remove the *Pinnaspis* species and thus infested fruit are not necessarily downgraded.

Figure 12. Cluster of males (top) and single female (upper left) of lesser snow scale, *Pinnaspis strachani*; and males with single female (bottom right) of fern scale, *Pinnaspis aspidistrae.*

YELLOW SCALE, *Aonidiella citrina* (Coquillett)

Yellow scale (Homoptera: Diaspididae), is an uncommon armored scale which can be found throughout the state, generally at low population densities. The scale cover of the adult female is very flat, circular, and orange-yellow in color (Fig. 13), and is much smaller than other circular armored scales such as Florida red scale, *Chrysomphalus aonidum* (L.). The armor is semitransparent and the yellow scale body can be seen through the scale cover. The scale body is kidney shaped and attached to the scale cover dorsally so that when the scale cover is turned over, the scale body is overturned as well. Immature yellow scale covers are similar to the adult cover but proportionally smaller. Male scale covers are elongate, smaller than adult female covers, and serve as a pupation chamber for the metamorphosing male. Adult males are typical winged fragile insects (Fig. 13) which live for a short period and mate with females located beneath the scale cover. Females reproduce by holding developing embryos internally until they are deposited beneath the scale cover as live yellow crawlers. Thus, no egg laying is associated with this scale species.

Yellow scale may be found primarily on foliage of citrus, but fruit and twigs also serve as feeding substrate. Woody bark is not a suitable substrate for yellow scale feeding. Some fruit spotting can be observed at yellow scale feeding sites. Only rarely do populations of yellow scale build to a point where they are of concern to producers. The basis for this low density existence may be a combination of marginal adaptation to environmental conditions present in Florida and the action of natural enemies.

Figure 13. Immature and adult female yellow scale on orange leaf (top); and two adult free living male yellow scale (bottom).

CHAFF SCALE, *Parlatoria pergandii* Comstock

Chaff scale (Homoptera: Diaspididae), is a trunk inhabiting armored scale which occurs widely in Florida and which is not considered a major pest problem. Its presence in many groves undoubtedly goes unnoticed due to its propensity for canopy interiors.

Adult female chaff scale are dull brown to gray, very flat and the armor cover is rather amorphous, taking on the shape of the area in which it develops (Fig. 14). On fruit, the scale tends to settle in dimples or depressions in the rind or under the calyx. On leaves, the midrib is favored, and the scales line up along the upper and lower midrib region (Fig. 14, top). On bark, the scales fill depressions and cracks. Successive generations congregate, and over time, the scale bodies build up adjacent to and on top of one another. This behavior and the amorphous shape lead to the common name chaff scale resembling grain chaff. Beneath the female scale armor is a circular flat female scale body which is purple in color. Reproductive females produce oblong purple eggs which are protected under the scale cover until hatching. Immature and male scale resemble the adult female but are smaller. Male scale covers tend to be more elongate than females and protect the developing winged adult scale as is typical of armored scales (Fig. 14).

Populations of chaff scale are slow to develop on citrus leaves, twigs, fruit and bark. With a developmental period longer than most armored scale and lower numbers of eggs per female, this scale may take years to reach high population densities. Feeding on leaves does not induce necrotic spots as occurs with purple scale, *Lepidosaphes beckii* (Newman), but fruit feeding does result in spotting, wherein the feeding sites do not color from green to yellow or orange at maturity (Fig. 14). Additionally, scale located in fruit depressions are difficult to remove from fruit during fresh fruit packing operations.

Field populations of chaff scale are held in check through the action of natural enemies, especially by the parasitic wasp *Aphytis hispanicus* (Mercet). Parasite induced mortality and slow rates of development keep this pest from becoming a perennial pest.

Figure 14. Adult females and immature male chaff scales on citrus leaf (top); and harvested grapefruit showing discolored areas where chaff scale has fed.

BLACK PARLATORIA SCALE, *Parlatoria ziziphi* (Lucas)

Black parlatoria scale (Homoptera: Diaspididae), is a relatively recent invader of Florida citrus, having been first detected in 1985 in the Miami area. The scale is still quite restricted in its distribution to residential citrus in the Miami area and has not yet been recorded from commercial citrus in the state.

Black parlatoria scale is very flat, and the armor of all stages is dull black (Fig. 15). Margins on reproductive adult females are light tan-gray. Adult female scale covers are elongate-oval (about 2 mm long) and only slightly convex, with the exoskeleton remains of the early nymphal molts evident at the perimeter of the armor. Beneath the scale cover, the adult female scale is purple. Eggs produced by black parlatoria and the hatching crawlers also are purple. Male scale armor is about one-third the size of the adult female, elongate, and black to brown in color.

Citrus leaves, fruit and twigs are infested by black parlatoria scale and adhesion of the scale to the plant substrate is extremely tight. Removal from infested fruit at the packinghouse is therefore impractical and infested fruit are generally rejected. Leaf and twig populations may build in some cases to the point where defoliation and twig dieback occurs. Populations of the scale can be found in all stages of development throughout the year, but development is relatively slow and females lay relatively few eggs (8 to 20). Thus, infestations neither build nor spread rapidly.

Some natural enemies have been identified in association with black parlatoria scale, but it is unclear how effective these natural enemies will be in maintaining populations below damaging levels.

Figure 15. Citrus leaf heavily infested with the black parlatoria scale, *Parlatoria ziziphi.*

BROWN SOFT SCALE, *Coccus hesperidum* L.

Brown soft scale (Homoptera: Coccidae), is widely distributed in Florida citrus groves, but generally is not an important problem. This species has one of the widest host plant ranges in the family coccidae, attacking countless fruit tree and ornamental plant species. The scale is flat and oval in shape during all life stages, and is light brown to yellowish in color (Fig. 16). The body of the scale becomes increasingly darker as the scale matures and a tougher exoskeleton develops. Mature female scale are 2.5 to 4 mm long, often mottled with brown spots (Fig. 16), but do not contain the dark internal ribbon characteristic of the closely related green scale, *Coccus viridis* (Green) (Fig. 17). Immature brown soft scale are not readily distinguishable from other related species, with which they may coexist. Female brown soft scale do not deposit eggs, but rather give birth to live young. Males are uncommon.

Brown soft scale populations are most often seen on leaves and twigs, but occasionally, fruit are infested. Populations of this scale tend to be highly aggregated, with many individuals found on a single leaf or stem. This is due in part to the high reproductive capacity of female scale. Three to 5 overlapping generations per year are reported on citrus.

High brown soft scale populations are accompanied by heavy deposits of honeydew, leading to considerable sooty mold coverage of foliage and fruit in the affected area (Fig. 16). Damage is most noticeable in young groves, where the combination of feeding and honeydew production can incapacitate and sometimes kill trees. Feeding on older trees may result in reduced tree vigor, twig dieback, and reduced yields, as well as fruit grade reduction. However, the presence of a rich complex of natural enemies in Florida is responsible for adequate biological control of brown soft scale under most circumstances, and only when this natural suppression is disturbed do scale populations become severe. Parasitized scale are commonly seen in most populations, and can be readily recognized by the darkened and swollen appearance of the affected scale (Fig. 16). Circular parasite emergence holes in various scale stages also are indicative of natural enemy presence.

Figure 16. Light female and immature of healthy brown soft scale and two dark parasitized scale (top left); citrus stem with colony of adult brown soft scale (top right); and citrus tree heavily infested with brown soft scale (bottom) showing heavy deposit of honeydew and growth of sooty mold fungi.

GREEN SCALE, *Coccus viridis* (Green)

Green scale (Homoptera: Coccidae), is closely related to brown soft scale, *Coccus hesperidum* L., and shares many physical and biological traits. All stages of green scale are oval and flattened, and in contrast to the tan-yellow color of brown soft scale, green scale are pale green in all stages (Fig. 17). Adult females are 1.5 to 3.7 mm in length and are distinguished from other similar soft scales on citrus by the highly visible black midgut ribbon which appears through the semitransparent dorsal body surface (Fig. 17). Additionally, black eye spots are quite visible along the anterior margins of the adult female. This species is functionally uniparental, that is, reproduction occurs through parthenogenesis. Males are absent or extremely rare.

Green scale further varies in reproductive strategy from brown soft scale in the deposition of eggs beneath the female scale. Brown soft scale gives birth to live young. Yellowish-green *C. viridis* eggs give way to tiny pale green crawlers which move from beneath the female scale and settle to feed on the leaf or green stem surface (Fig. 17). Immature stages of green scale are indistinguishable from closely related species, and to further complicate matters, often are found in mixed populations with other soft scale species.

Green scale is primarily restricted to new flush growth, and as a result, often is visible on outer tree canopies. Spotty sooty mold patches develop on foliage beneath green scale colonies where honeydew from scale insects accumulates, and this is generally the first indication of green scale infestation. Leaf midribs and green stems are favored feeding sites (Fig. 17), but as populations increase, scale insects spread out on leaves and can be found on both surfaces of the leaf as well as occasionally on fruit. Populations are more synchronized than many other scale species on citrus, and become visible on spring flush and gradually increase during summer months. Parasites, predators and fungi are responsible for high mortality of immature and adult scale, and fungal disease spread brings about dramatic population crashes during wet periods in late summer and fall.

Figure 17. Adult female green scale, *Coccus viridis* and crawler (top); and immatures of green scale along leaf midrib (bottom).

BLACK SCALE, *Saissetia* spp.

The black scale (Homoptera: Coccidae), on citrus in Florida historically has been a source of confusion. Early reports of the scale referred to *Saissetia oleae* (Bern.) and later it was discovered that the species most often encountered in Florida citrus was actually the Caribbean black scale, *Saissetia neglecta* DeLotto. The following discussion relates to Caribbean black scale, but also is generally applicable to the two closely related but lesser common species.

Mature female scale are necessary for diagnosis, as is the case for most other scale insect species. The female is dark brown-black, hemispherical in shape, and is distinguished from other scales on citrus by the appearance of a series of ridges on the dorsal surface which make an H pattern on the scale (Fig. 18). The raised areas are less developed but generally visible in immature stages as well. Female black scale are nearly round, approximately 1.5 to 4.5 mm long, 1.0 to 4.0 mm wide, and very convex. Mature females deposit up to 4,000 pink eggs beneath their body. Light brown crawlers move from beneath their mother to begin feeding on leaf or twig surfaces. Young scales are very flat and oval in shape, and are relatively indistinguishable from closely related species until the conspicuous H pattern begins forming in the second instar (Fig. 18). In the third instar, the scale becomes more convex and darkens slightly in color while the exoskeleton remains soft. This "rubber stage" transforms later to the mature female through an additional molt, after which the exoskeleton darkens and becomes more rigid. Initial feeding sites for immature black scale are leaf, twig and fruit surfaces. As the scale develops and molts, movement to twigs occurs, so that most adult females are located on twigs. Fruit stems and twigs near fruit are favored sites. Coupled with prolific secretion of honeydew, this movement results in location of sooty mold where fruit quality will be affected. Leaf and fruit drop may also occur, and twig dieback has been observed in conjunction with sustained feeding by black scale. Generally 3 to 4 somewhat synchronized generations are completed per year in Florida, with the synchronization becoming weaker in the south. By and large, the pest is under satisfactory biological control by natural enemies, particularly the wasp, *Scutellista cyanea,* which acts as an egg predator.

Figure 18. Life stages of black scale, *Saissetia neglecta* on citrus stem. Nymphal stages showing characteristic "H" ridges on top of body (top); and adult female scales on citrus branch (bottom).

HEMISPHERICAL SCALE, *Saissetia coffeae* (Walker)

The hemispherical scale (Homoptera: Coccidae), is named for the highly convex, rounded shape of the female scale (Fig. 19). Closely related to the black scales, *Saissetia* spp., hemispherical scale is very similar in appearance and habits. The female scale is 1.5 to 3.5 mm in diameter and has a smooth shiny exoskeleton without the noticeable ridges characteristic of black scale. The adult female color is yellowish-brown to dark brown, and has a lighter brown margin. Immature stages are lighter in color, with pinkish nymphs emerging from eggs and transforming to tan coloration shortly thereafter. Immature stages exhibit the ridged H pattern symptomatic of black scale immatures, and thus are easily confused with black scale unless adult females are present.

Hemispherical scale shares habits with the black scales, attacking leaves and young twigs as immatures. Movement to older wood occurs as nymphs progress to adult females, but to a lesser extent than with black scale. As a result, adult females are found on leaf tissue as well as on twigs and branches and occasionally on fruit. Eggs are laid beneath the female body in a cavity, and as such are susceptible to attack by the predatory wasp, *Scutellista cyanea* Motschulsky. Two or more generations are completed per year, with gravid females being the stage commonly encountered in winter. Overlapping generations occur in warmer parts of the insects' range.

Hemispherical scale is generally considered to be under adequate biological control through the action of *S. cyanae* and a group of parasites, and thus is not a serious pest on citrus in Florida. Other host plants, including cycad, are more intensively infested by the scale, and supplemental control may be required on these plants.

Figure 19. Adult female hemispherical scale on citrus leaf.

PYRIFORM SCALE, *Protopulvinaria pyriformis*

Pyriform scale (Homoptera: Coccidae), is an uncommon pest of citrus and several other plant species in Florida. The name is derived from the pear-shaped body of the scale, particularly the mature female. This scale is distinct from other scale insects found on citrus in several characteristics besides the shape of the female. The scale is quite flat in all stages and the female develops a white cottony wax fringe that is evident at reproductive maturity (Fig. 20). The scale body color varies from light green in early stages to tan, brown, or reddish-brown as the scale matures (Fig. 20). The darker color of adults does not extend to the margins, leaving a lighter halo around the exoskeleton, which is further accentuated by the white wax fringe. Eggs are deposited within the waxy covering beneath the female, where they are protected from the environment. Hatching crawlers are light green in color and settle in the vicinity of their mothers. Successive immature stages also are light green in color and can be easily confused with other scale species, including green scale, *Coccus viridis* (Green). The pear shape of pyriform scale is most noticeable in the adult female, but also is visible in the immature stages.

Pyriform scale is generally restricted to leaves, and inhabits the lower surfaces almost exclusively. Localized patches may develop unnoticed, and occasionally build to the level where sooty mold is visible in association with honeydew secreted by the scale. Several overlapping generations are completed per year, but rarely have populations been considered of economic consequence on citrus. Higher population levels have been observed on ornamental plants such as gardenia. Several parasite species have been observed attacking pyriform scale, and the action of these natural enemies may be partially responsible for their limited pest status.

Figure 20. Pyriform scale, *Protopulvinaria pyriformis* on citrus. Immature scales (top); and reproducing females showing waxy fringe containing eggs (bottom).

FLORIDA WAX SCALE, *Ceroplastes floridensis* Comstock and BARNACLE SCALE, *C. cirripediformis* Comstock

Several species of wax scales are found on Florida citrus, but generally they occur in low numbers due to the presence of effective natural enemies. Wax scales (Homoptera: Coccidae), are distinguished by the presence of thick wax covering the exoskeleton in immature and adult stages. Species vary in the shape, sculpting, and color of the waxy covering as described below. Wax scales are generally found during the immature stages on leaves and twigs, and mature females are most often encountered on twigs and branches, where they are often overlooked. Leaf populations often are aligned with the midrib. The pink or dark red eggs are protected beneath the female scale much the same as hemispherical scale and black scale, and hatching crawlers are generally pink in color and move from beneath the female to settle and begin extracting sap from leaves or twigs. Like many of the other species of soft scales encountered on citrus, wax scales secrete generous amounts of honeydew on which sooty mold develops, and often the presence of sooty mold is the first indication of infestation by this group of scales.

The Florida wax scale, *Ceroplastes floridensis* Comstock, is the most common species on Florida citrus. This species is distinguished by its dirty white-pink coloration and lack of visible plates or horns on the waxy surface of the scale body. Spiracular wax bands are visible. The adult female scale is 2 to 4 mm in length and 1 to 3.5 mm in width. Immatures are flat and may bear waxy projections laterally from the body margins (Fig. 21), while adult females are oval in shape and strongly convex, almost hemispherical (Fig. 21). Reproduction is parthenogenic in this species, and males are unknown.

The barnacle scale, *Ceroplastes cirripediformis* Comstock is similar to the Florida wax scale, but is distinct from the latter by the presence of 6 marginal and 1 middorsal plates in the adult female (Fig. 21). Adult females are 1 to 7 mm long and 1 to 5 mm wide. Immatures are similar in appearance to those of Florida wax scale. Winged males of this species are known but rare.

Figure 21. *Ceroplastes* wax scales. Young immatures of Ceroplastes near citrus leaf midrib (top left); adult female barnacle scale, *Ceroplastes cirripediformis* (top right); 2nd instars of Florida wax scale, *Ceroplastes floridensis* (bottom left); and late nymphal stage and adult female Florida wax scale on stem (bottom right).

RED WAX SCALE, *Ceroplastes rubens* Maskell and STELLATE SCALE, *Vinsonia stellifera* (Westwood)

The red wax scale, *Ceroplastes rubens* Maskell (Homoptera: Coccidae), has been reported in Florida since the 1950's, although its current status is unknown. This species is 2 to 5 mm in length and 1 to 4 mm in width, and somewhat pentagonal in dorsal view. Curved wax bands are evident at the points of the pentagon, as is a central depression (Fig. 22). The waxy covering of the scale is pink to reddish brown in all stages.

A fourth species of wax scale, the stellate scale, *Vinsonia stellifera* is quite distinct from the *Ceroplastes* species discussed above. Immatures and adult females of this species are pink-purple and star shaped (Fig. 22). The rays of the star number 5 or 6, and are composed of semitransparent wax. The adult female is similar in size to *Ceroplastes* spp. found on citrus and has similar habits in immature and adult stages. This species has previously been eradicated from Florida and its present status is uncertain.

Wax scales overwinter as females in central to northern portions of the state, but tend to remain active year around in southern portions. In addition to citrus, numerous ornamentals are infested by these scales. The presence of several effective natural enemies, the most notable of which is the egg predator, *Scutellista cyanea* Motschulsky, helps to maintain the status of these wax scale species as occasional pests on citrus.

Figure 22. Wax scales (Continued). Red wax scale on leaf (top); and immature of stellate scale, *Vinsonia stellifera* (bottom). Note 1st instar nymph of stellate scale (bottom, lower left corner).

COTTONYCUSHION SCALE, *Icerya purchasi* Maskell

Cottonycushion scale (Homoptera: Margarodidae), has been reported in Florida since the 1890's, and is easily distinguished from other scale insects on citrus. The mature female scale body is broadly oval, bright orange-red, yellow or brown, but is largely obscured by a large fluted egg sac which protrudes from under the posterior of the mature female body (Fig. 23). This egg sac is made of white or yellowish wax and is about twice as long as the scale body proper, giving the female scale an overall length of about 15 mm. The egg sac contains 500 to 800 reddish eggs which hatch in a few days during warmer temperatures and up to 2 months during cooler periods. Newly hatched nymphs are bright red with dark antennae, and settle upon small twigs and leaves (Fig. 23). Third instar scale are covered with a thick cottony secretion which subsequently disappears after the third molt (Fig. 23).

About 4 months is required to complete a generation in this species. Adult female scale are actually hermaphroditic, that is, they are capable of self fertilization. Presence of males is not necessary for reproduction, and in general, the dark red, winged males are rare.

Early stages of the scale extract plant sap while feeding along leaf midribs and green twigs, and this early feeding causes the greatest damage. Third instars of the scale migrate to larger twigs, where they feed and secrete honeydew upon which sooty mold can grow. Adult females generally are found on larger branches or the trunk, and can build up large encrustations in the absence of natural enemies (Fig. 23). Fruit are rarely attacked by this scale in Florida. Heavy infestations occasionally may build in mature trees, resets, and in nurseries, and these populations can lead to defoliation, twig dieback, and fruit drop. Early stages are most responsible for dispersal of the insect, as the young nymphs are wind blown to other plants or crawl from plant to plant. In addition to citrus, this insect attacks numerous ornamental plants. Cottonycushion scale is seldom a serious problem due to successful introduction of the ladybeetle, *Rodolia cardinalis* (Mulsant) around the turn of the century. Larvae and adults of the Vedalia beetle devour cottonycushion scale and usually keep the scale from reaching economic levels.

Figure 23. Cottonycushion scale, *Icerya purchasi.* Immature scale turned over to show legs and antenna (top left); 2nd and 3rd instars on leaf (top right); heavy infestation on pittosporum shrub (bottom left); and adult female cottonycushion scale with expanded waxy egg sac (bottom right).

CITRUS MEALYBUG, *Planococcus citri* (Risso)

The citrus mealybug (Homoptera: Pseudococcidae), is a pest most often encountered in sheltered locations on the citrus tree, such as cracks, crevices, and folds in leaves and bark, and in fruit clusters or under fruit buttons. Easily recognized by visible segmentation on the body and powdery wax covering, mealybugs remain mobile throughout their life. The soft body of citrus mealybugs is oval and yellowish in color. Wax secretions cover the body to varying degrees, with newly hatched nymphs being virtually free of wax and other stages lightly to heavily covered in fine powdery wax (Fig. 24). Mated female mealybugs are about 3 mm long and produce 300-600 eggs which are covered in a cottony mass of wax (Fig. 24). These eggs require from 6 days to several weeks to hatch. Newly hatched nymphs are light yellow and progress through 3 immature stages, after which males pupate to ultimately become winged gnat-like adults typical of scale insects. Nymphal females molt to an adult stage which retains the nymphal form.

Populations of citrus mealybug complete 2 or 3 overlapping generations per year in Florida, and appear to prefer certain cultivars. Grapefruit are more heavily infested than oranges, and clusters of grapefruit often shelter mealybug populations. Mealybugs, like scale insects, extract sap from the tree while feeding and excrete large amounts of honeydew.

The citrus mealybug is the only mealybug species important to Florida citrus, and generally occurs at low levels due to the combined actions of a ladybeetle predator, often called the mealybug destroyer, *Cryptolaemus montrouzieri* (Mulsant), several parasitic wasps, and a fungal disease which act in concert to provide biological control. Localized populations occasionally reach high levels, resulting in damage. Severe infestations can lead to defoliation, and feeding of mealybugs under fruit calyces or on fruit stems may result in fruit drop. Citrus mealybug feeding on fruit surfaces may result in discoloration, bumpy texture and scarring, lowering fruit grade. While reduced yields can result from this feeding, the greatest damage by citrus mealybug occurs as a result of heavy honeydew and sooty mold associated with the pest, particularly in fresh fruit production.

Figure 24. Citrus mealybug colony on grapefruit cluster (top); and adult female citrus mealybug (bottom left) with wax egg sac (right).

CITRUS WHITEFLY, *Dialeurodes citri* (Ashmead)

Citrus whitefly (Homoptera: Aleyrodidae), preferentially attacks citrus, but also infests such ornamental plants as gardenia and ligustrum. Adults of this whitefly species, as with all whiteflies, are minute (1 to 2 mm long) winged moth-like insects (Fig. 25). Adults are covered with a white waxy powder and bear no distinguishing marks or pigmentation on the adult body. The abdomen is yellow and the wings are held roof-like over the body. Males are generally slightly smaller that females and all stages of citrus whitefly are smaller than corresponding stages of its congener, the cloudy-winged whitefly, *Dialeurodes citrifolii* (Morgan). Mated adult female citrus whiteflies lay about 150 eggs more or less at random on the underside of leaves (Fig. 25). Immature leaves are preferred egg laying (oviposition) sites, and interior water sprouts are particularly likely to become infested by this species. The yellowish unsculptured eggs are anchored to the leaf surface by a short stalk. Emerging crawlers are yellow and, after settling on the underside of a suitable leaf, they begin feeding by inserting their mouthparts into the leaf tissues. From this point onward, immature citrus whiteflies are sessile. Nymphs are oval, flat and clear, making them extremely difficult to see when populations are light (Fig. 25). Eyespots and developing wing pads are often visible through the immature integument of final nymphal stages, as are developing internal parasite immatures when attack by these natural enemies occurs. Generation times range upward from about 45 days.

Populations of citrus whitefly build on spring and summer flushes, generally increasing in intensity as the season progresses. Moderate to high citrus whitefly densities are evident by localized buildup of sooty mold on honeydew excreted by the citrus whitefly. Under moist summer conditions,citrus whitefly often is infected by the fungus, *Aschersonia aleyrodis* Webber, which may spread rapidly through dense citrus whitefly populations. Resulting epidemics cause whitefly collapse, and this fungus has been credited with successful biological control of citrus whitefly since early in this century. Additional mortality induced by predators and parasites appears to keep citrus whitefly in check in commercial citrus.

Figure 25. Adult and eggs of citrus whitefly on underside of leaf (top); and mature nymph (so-called pupa) of citrus whitefly from which adult will emerge (bottom).

CLOUDY-WINGED WHITEFLY, *Dialeurodes citrifolii* (Morgan)

The cloudy-winged whitefly (Homoptera: Aleyrodidae), is closely related to the citrus whitefly, *Dialeurodes citri* (Ashmead) and as such, shares many physical and biological traits with this allied species. Adults of the cloudy-winged whitefly are noticeably larger than citrus whitefly adults, and often can be characterized in the field by the presence of lightly shaded or clouded areas toward the posterior of the wings (Fig. 26). Adults congregate on young citrus leaves, where they lay yellow elongate-oval stalked eggs very similar in appearance to those of citrus whitefly. Coloration of the eggs darkens as the eggs mature, allowing for some indication of the age of eggs on citrus foliage. These eggs are generally found on the leaf undersides, and their preference for laying eggs on young succulent citrus leaves results in active populations of cloudy-winged whitefly (and citrus whitefly) being focused on citrus terminal growth and on interior succulent sprouts. Nymphs of cloudy-winged whitefly are clear and extremely flat, and like the nymphs of citrus whitefly, are quite difficult to see (Fig. 26). In the fourth nymphal instar, known as the pupa, the appearance of red eye spots and light colored wing buds increases the visibility of the immature cloudy-winged whitefly. A hand lens is generally required to locate immature whitefly on the leaf surface.

Populations of cloudy-winged whitefly are seasonally tied to tree flushing patterns, and like citrus whitefly, populations of this species increase through the summer months and are most evident by the successive buildup of honeydew and sooty mold on leaf surfaces surrounding the infestation. Severe populations can turn the outer canopy of infested trees quite black. Mixed populations of cloudy-winged and citrus whitefly often can be found under these high population situations. Both citrus whitefly and cloudy-winged whitefly are susceptible to the naturally occurring fungus, *Aschersonia aleyrodis* (Webber) (Fig. 26) and other less common fungal species. In contrast, cloudy-winged whitefly is not readily attacked by the aphelinid wasp *Encarsia lahorensis* (Howard) which contributes substantially to citrus whitefly population management. There apparently are no effective parasite species which assist in cloudy-winged whitefly biological control.

Figure 26. Adult and pupal skin of cloudy-winged whitefly on underside of citrus leaf (top); several nymphs of cloudy-winged whitefly on leaf (bottom left); and leaf infestation (bottom right) of cloudy-winged whitefly killed by the fungus *Aschersonia aleyrodis* (Courtesy R. L. Phillips). White halos with red centers are carcasses of whitefly from which fungus has erupted and produced fruiting bodies.

WOOLLY WHITEFLY, *Aleurothrixus floccosus* (Maskell)

Woolly whitefly (Homoptera: Aleyrodidae), is commonly encountered in Florida citrus, but rarely reaches damaging populations. The whitefly occurs in close nymphal patches on leaf undersides, and in higher populations, honeydew and sooty molds are evident.

Adults of woolly whitefly are similar to other whitefly species on citrus, with yellowish coloration and unmarked wings. Adults lay eggs in clusters on mature leaves (Fig. 27, top), and these eggs are cylindrical, beige in color, and attached to the leaf by a short pedicel. Fine powdery wax is often laid down by female wooly whitefly in the vicinity of the eggs.

Nymphs emerging from eggs generally settle near the eggs and thus form a cluster (Fig. 27). Nymphs are light green in color, and progressively darken to tan as they develop and molt. The final nymphal stage, the pupa, is brown in color but is covered completely with fine waxy threads (Fig. 27). These waxy threads create a woolly appearance which gives the whitefly its name, but the threads differ from thicker wax rods produced from dorsal pores in nymphs of nesting whiteflies. Woolly whitefly lacks dorsal wax pores and an additional distinction is that the majority of wax threads are produced by woolly whitefly only in the pupal stage.

Feeding by woolly whitefly is typical of other whitefly species, with adults and nymphs extracting plant sap from leaf tissues and secreting the excess as honeydew. High population densities affect plant growth through the promotion of sooty mold (Fig. 27), and this condition can be seen in association with numerous other homopterans. Under most conditions, woolly whitefly in Florida citrus is regulated by natural enemies, including parasitic wasps, predators, such as ladybeetles, and fungal disease. Typically this whitefly is observed in scattered patches and is most visible in the canopy interior.

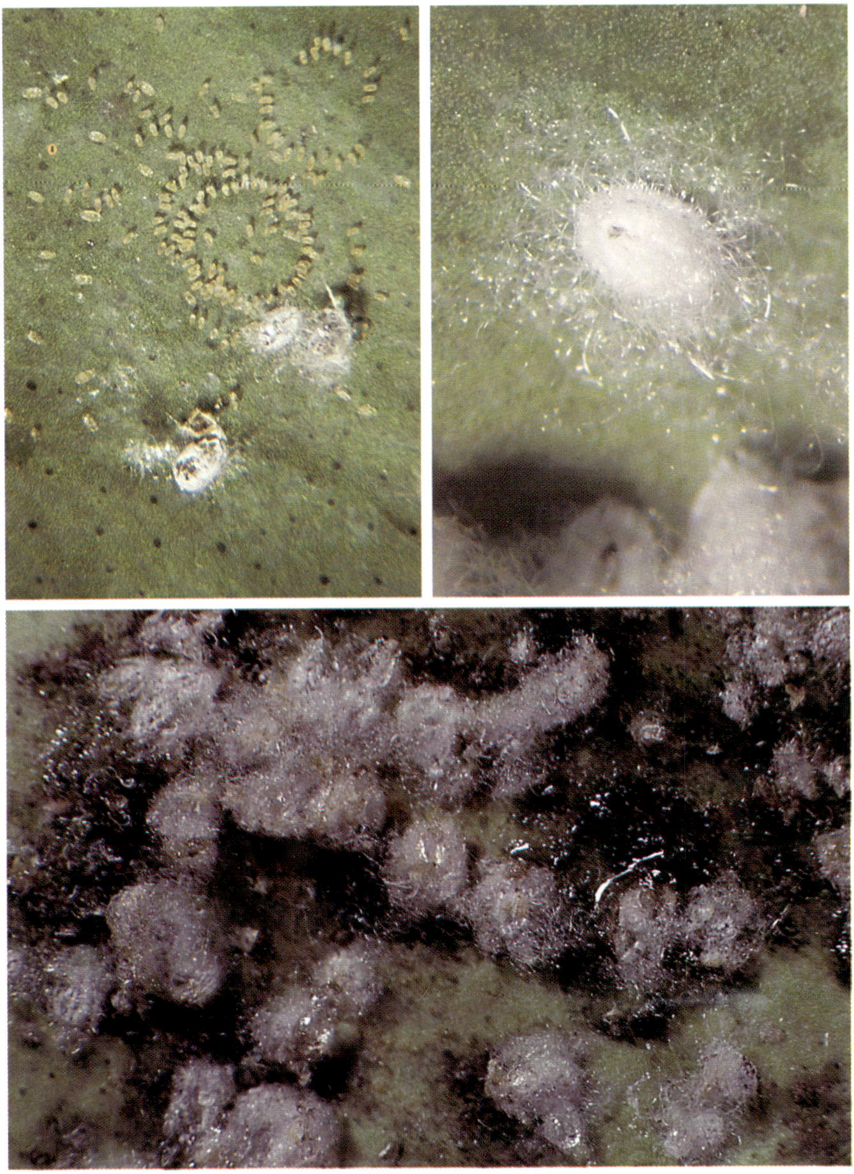

Figure 27. Egg masses and nymphs of woolly whitefly, *Aleurothrixus floccosus* on citrus leaf (top left); mature nymph of woolly whitefly with dense cottony wax covering (top right); and colony of woolly whitefly (bottom), showing honeydew pools and sooty mold developing.

NESTING WHITEFLIES, *Paraleyrodes* spp.

Among the aleyrodids attacking Florida citrus are the nesting whiteflies (Homoptera: Aleyrodidae), belonging to the genus *Paraleyrodes.* Probably representing more than one species, these whiteflies are distinguished from other species on citrus by the habit of adult females of constructing a wax "nest" in or around which her eggs are laid. Her nest often incorporates the cast skins (exuviae) and associated wax threads of immature whiteflies which are present on the leaf (Fig. 28, top).

Adult *Paraleyrodes minei* are similar in appearance to other whiteflies, but there are distinguishing features for the genus. The antennae are generally held forward in a bent position, a characteristic not observed in other citrus whiteflies. Wings of adult nesting whiteflies are held rather flat (Fig. 28) as opposed to the inverted V formation of the similar appearing woolly whitefly, *Aleurothrixus floccosus* (Maskell). Some wing markings may occur on adult nesting whiteflies, but this marking is not consistent nor is it diagnostic.

Eggs of Paraleyrodes spp. are typical of whiteflies, being elongate cylindrical and having a short pedicel with which the egg is anchored to the leaf tissue (Fig. 28). Egg number per nest varies from just a few to as many as 30 or 40 eggs. Nymphs hatching from these eggs disperse over the leaf and onto adjacent leaves before settling to feed, and thus clusters of developing nymphs are not normally encountered as they are with the woolly whitefly. As nymphal development ensues, a wax margin of flat bands develops around the perimeter of the nymph and numerous large crystalline wax rods emanate from wax pores on the dorsum of the nymph (Fig. 28). As development proceeds, the wax rods elongate and break off, littering the leaf surface with the threads which make the nymph quite noticeable. Development occurs almost exclusively on leaf undersides.

Populations of nesting whiteflies in Florida represent *P. minei* and at least one other species. These whiteflies are generally widespread in the state, but occur at relatively low numbers. Several natural enemy species (parasitic wasps and ladybeetles) are associated with nesting whiteflies in Florida, and it is assumed they are in part responsible for the relatively low populations generally encountered despite the year-around occurrence of the species.

Figure 28. Nesting whitefly, *Paraleyrodes minei* adult on nest, eggs and various nymphs on leaf underside (top); and mature nymph of nesting whitefly, highlighting the thick wax rods characteristic of this species (bottom).

BAYBERRY WHITEFLY, *Parabemisia myricae* (Kuwana)

A whitefly species relatively new to Florida is the bayberry whitefly (Homoptera: Aleyrodidae), first recorded in 1983 on ornamental shrubs in south Florida. This whitefly is smaller than other species in Florida citrus, and as such is less noticeable at low population levels. However, tremendous populations build in which render other new infested citrus producing areas growth unsightly with honeydew and sooty mold growth.

The adult bayberry whitefly is about half the size of citrus whitefly, *Dialeurodes citri* (Ashmead), is pure white, and often congregates on new citrus growth (Fig. 29). Small emerging leaves (feather growth) are the site of egg laying (oviposition) by the bayberry whitefly, and this species oviposits on younger growth than other species.

Eggs of bayberry whitefly are dark in color and are attached to leaf margins and both upper and lower leaf surfaces by a short pedicel. This apparent indiscriminate use of leaf surfaces is a product of their use of extremely young, succulent leaf tissue for oviposition. Hatching nymphs are extremely small and settle to feed on the young tissue as do other whitefly species. The nymph is greenish-yellow in color and has no evident adornment with wax. Nymphs of all instars are oval and convex, with red compound eyespots becoming visible as nymphal maturity occurs and the "pupal" stage progresses (Fig. 29).

The life cycle of bayberry whitefly is completed in 30 days or less under ideal conditions, so populations can develop quickly. Numerous overlapping generations occur per year, so that infestation can be expected whenever flushing occurs in infested areas.

Bayberry whitefly has not yet become a serious pest in Florida citrus despite its presence in the state for some 10 years. Severe infestations have been observed in a few citrus nurseries (Fig. 29), but limited infestations have been encountered in mature commercial citrus groves. The presence of parasites may assist in keeping bayberry whitefly populations below damaging levels in commercial citrus.

Figure 29. Adult and mature nymphal skin of bayberry whitefly, *Parabemisia myricae* on young leaf (top left); mature nymph (pupa) showing red eye spots (top right); and trifoliate citrus nursery seedlings showing sooty mold associated with bayberry whitefly infestations (bottom).

CITRUS BLACKFLY, *Aleurocanthus woglumi* (Ashmead)

Since its introduction and permanent establishment in Florida in 1976, this species of whitefly (Homoptera: Aleyrodidae), has caused severe damage in dooryard and some commercial citrus groves (Fig. 30). Extremely high populations on mature trees have resulted in overall blackening of the foliage through development of sooty mold fungi on the honeydew secreted by blackfly nymphs, and this shading has resulted in stunted tree growth, reduced fruit size, and limited bloom in subsequent years.

Citrus blackfly is distinct from other aleyrodids on citrus in that all life stages except the egg are black. Eggs are yellow-orange when freshly laid, and they occur in a spiral configuration of up to 50 eggs on leaf undersides (Fig. 30), characteristic of citrus blackfly and other *Aleurocanthus* species. Nymphs hatching from the darkened eggs tend to settle in the vicinity of the egg cluster or young as well as mature leaves. The molt to second nymphal instar yields a slightly longer, broader black insect, and subsequent nymphal instars are progressively more oval and convex in shape. The presence of large bristles (setae) on the dorsal and lateral surfaces of blackfly nymphs also separates them from other citrus-infesting whiteflies in Florida. Cast skins from previous molts may be impaled on the bristles of later nymphal instars. A white waxy margin appears in the fourth nymphal instar (also called pupa), and this margin is useful in determining the sex of pupae (Fig. 30). Male pupae are smaller than females and tend to have a more complete margin of wax compared to the larger female pupae. Emergent adult blackfly initially are bright red, but the adults become slate black in color and wings, held roof-like over the body (Fig. 30). Adult blackflies often are seen in numbers over fresh citrus leaf growth (Fig. 30).

The introduction of 2 non-indigenous parasites, *Amitus hesperidum* (Silvestri) and *Encarsia opulenta* (Silversti) has led to successful biological control of the pest. Under extremely high populations, blackfly has been observed to lay eggs on a broad range of shrubs and trees, but few if any of these other hosts can support growth and development of the insect.

Figure 30. Citrus blackfly, *Aleurocanthus woglumi.* Leaf infestation of egg spirals and mature nymphs of blackfly (top left); grapefruit on tree heavily infested with citrus blackfly and covered with sooty mold (top right); adult blackfly on leaf surface (bottom left); and young citrus terminal covered with adult blackfly (bottom right).

MELON APHID, *Aphis gossypii* Glover

The melon aphid (Homoptera: Aphidae), is a relatively common aphid species in Florida citrus, and like all citrus aphids, is closely associated with new growth flushes and emergent succulent leaves. This aphid is grey-green in color when feeding on citrus and is quite small (less than 2 mm) relative to other aphid species on this host (Fig. 31). In comparison, the black citrus aphid, *Toxoptera aurantii* (Boyer de Fonscolombe), is shiny black (Fig. 33) and the green citrus aphid, *Aphis spireacola* Patch, is light yellow-green on citrus (Fig. 32).

Female winged forms (alates) colonize new terminals and young leaves, and begin to produce live young after just a few days (Fig. 31). Rapid population growth is initiated with spring flush, and generally continues until elevated temperatures, gaps in leaf flush and natural enemies reduce aphid survival and reproduction. Only 3 days are required for immature development so that generations advance very quickly. In addition, this aphid produces no males when on citrus and thus experiences no developmental delay due to mating and egg maturation. Generations of melon aphids following initial colonization are composed predominantly or exclusively of wingless individuals.

Large colonies develop as a result of short generation time and up to 60 or so progeny per female. Within a few weeks of the initial infestation by a single female, an infested terminal may contain tens or hundreds of aphids, resulting in distorted growth as evidenced by leaf cupping and twisting. The result of this damage is reduced photosynthetic capability and thus tree growth may be affected. As with other foliar pests, melon aphid is more damaging to small trees in the field and in the nursery than it is to larger trees, and seldom reaches populations which cause economic injury. Parasitic wasps and predators such as ladybeetles, lacewings and syrphid flies are in part responsible for maintaining melon aphids at low levels.

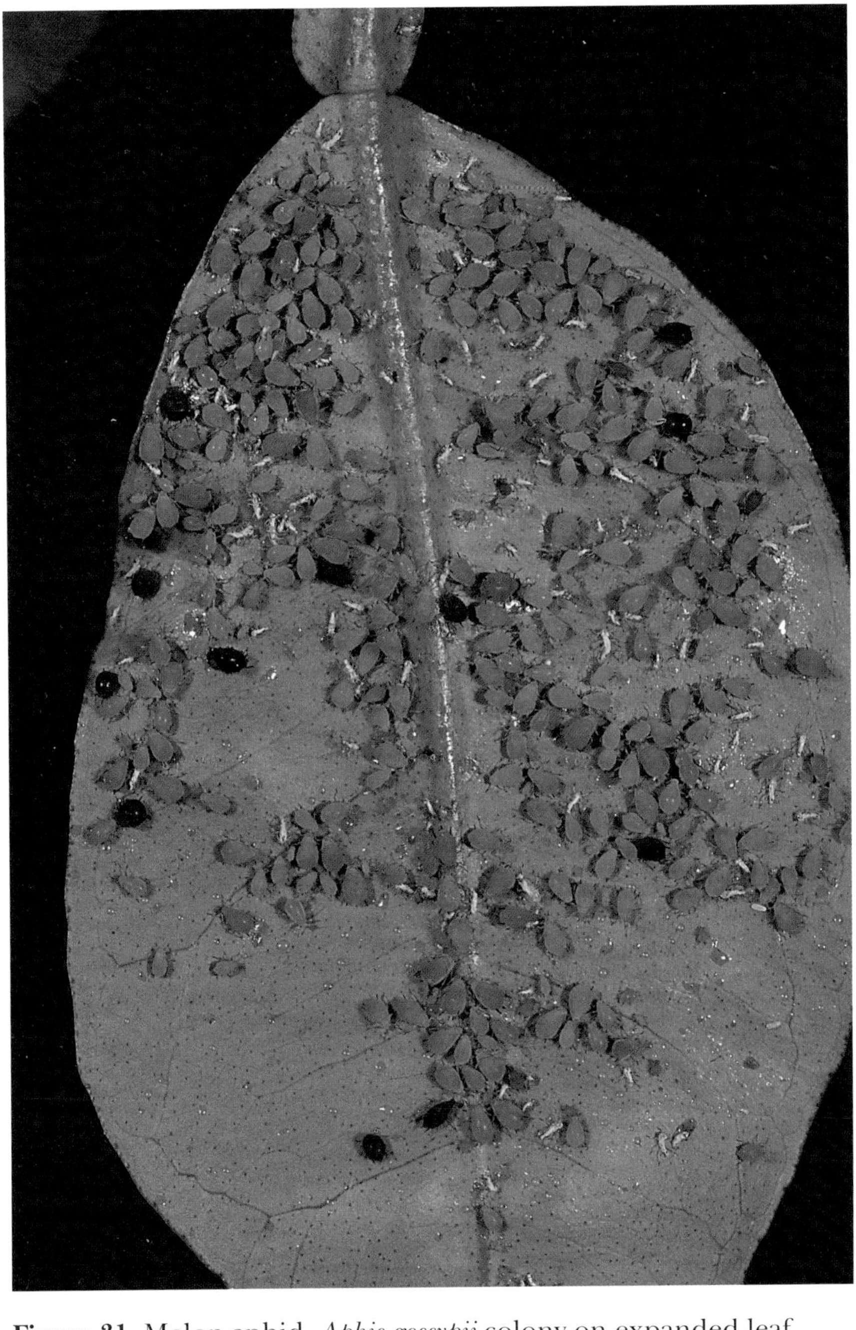

Figure 31. Melon aphid, *Aphis gossypii* colony on expanded leaf.

GREEN CITRUS APHID (SPIREA APHID), *Aphis spireacola* Patch

The green citrus aphid (Homoptera: Aphidae), is a common aphid species on citrus, and is probably the most routinely encountered aphid across citrus groves in Florida. It also is characterized as being the lightest in coloration (Fig. 32) and thus is not normally confused with the other species, melon aphid, *Aphis gossypii* Glover, and black citrus aphid, *Toxoptera aurantii* (Boyer de Fonscolombe). Populations of green citrus aphid have been recorded statewide on citrus and ornamental shrubs since early in the century.

Colonization of new sites occurs through flights of winged forms (alates). However, typical of other citrus aphids, seasonal populations are composed primarily of live-bearing (viviparous) females which do not possess wings (apterous) and which reproduce quickly. These females are 1 to 2 mm long, light green-yellow in color, and are encountered on citrus terminals where they infest young leaves. Females produce offspring soon after they reach maturity and this feature, coupled with relatively high reproductive rates, brings about the potential for rapid population buildup. A mixed population of all stages results and cast exoskeletons are observed scattered in the area of infestation (Fig. 32). Winged adults have dark head and thoracic regions with green abdomens (Fig. 32).

New citrus flush is vulnerable to attack by this and other aphid species, and sustained feeding results in leaf curling and stunting (Fig. 32). Honeydew production by these and other aphids can be observed on infested flush, and sooty mold fungi may build up around the infestation as well. Populations are most noticeable at spring flush and early summer flushes, but tend to decline as summer temperatures are experienced. The decline appears to be due to the combination of fewer new leaves, higher temperatures, and the action of natural enemies, including parasites, predators, and pathogens. As with other aphids on Florida citrus, high populations of green citrus aphid attract generalist predators such as coccinellids, which lay eggs in aphid colonies. Hatching predator larvae begin consuming aphids immediately, but decline of aphid populations usually doesn't occur until after some leaf distortion has occurred.

Figure 32. Green citrus aphid, *Aphis spireacola* colony on citrus leaf (top left), including numerous cast skins (white); winged adult (top right); and leaf curling and distortion resulting from feeding by green citrus aphid (bottom).

BLACK CITRUS APHID, *Toxoptera aurantii* (Boyer de Fonscolombe)

Black citrus aphid (Homoptera: Aphidae), is the least commonly encountered of the three aphid species on citrus in Florida, and unlike the melon aphid, *Aphis gossypii* Glover, and green citrus aphid, *Aphis spireacola* Patch, is more adapted to feeding on older leaves. Large colonies on terminals typical of the other species are replaced with smaller colonies of black citrus aphids feeding on undersides of expanded leaves. Black citrus aphid is dark reddish brown to shiny black in appearance in contrast to the blue-green of melon aphids. Wingless forms (apterous) are about 2 mm long, and thus are the largest of the Florida citrus aphids. They are reproductive females which produce live young parthenogenically (Fig. 33), a characteristic of all three aphid species in Florida. During the majority of the citrus growing season, populations of black citrus aphid are composed of this wingless form. However, in spring when new colonies are being established, or later, when colonies on terminals are being crowded, winged forms (alates) may be present also. The winged black citrus aphids are similar in appearance to wingless forms, being shiny black in color and quite large (Fig. 33). The wings show a dark spot along the front margin, and the median vein only exhibits one branch. This is a distinguishing feature of this species from the closely related brown citrus aphid, *Toxoptera citricida* (Kirkaldy), a serious citrus pest not yet reported from Florida.

Black citrus aphid is widespread in distribution in citrus producing regions of the world, and also is known to attack Camellia. It is a weak vector of the citrus tristeza virus, whereas, the related brown citrus aphid is an efficient vector of the virus and therefore is a continuing threat to the citrus industry of Florida.

Figure 33. Black citrus aphid, *Toxoptera aurantii* (top), wingless adults and immature aphids on young leaf; and winged adult black citrus aphid on leaf (bottom).

BROWN CITRUS APHID, *Toxoptera citricida* (Kirkaldy)

The brown citrus aphid (Homoptera: Aphidae), is native to Asia, and as of this writing, has not been reported from Florida or from the continental United States. Concern over the introduction of this aphid is based on its status as a principal vector of tristeza virus of citrus. Brown citrus aphid is closely related to and thus very similar in appearance and habits to the black citrus aphid. Adults are similar to other dark aphids, but there are some characteristics which distinguish brown citrus aphid from the closely related black citrus aphid. First, and probably most definitive, is the appearance of a twice-forked medial vein on the wing in the winged form (alate) of brown citrus aphid. This second or terminal fork is visible in live winged aphids with the aid of a handlens, but is missing from the wing of the black citrus aphids. Additionally, the third (basal) segment of the antenna of brown citrus aphid is black, compared to adjacent segments, whereas in black citrus aphid, this segment is clear with a black tip. The antennal character is visible in both winged and wingless forms of brown citrus aphid.

Brown citrus aphid is relatively limited in host range, preferring citrus and related species of Rutaceae. All stages of the aphid feed on leaf and green twig material (Fig. 34) by extracting phloem and depositing honeydew on plant surfaces adjacent to the colony and dense colonies may develop. Piercing of the plant tissue with the mouthparts is responsible for acquisition of the tristeza virus from infected plants and subsequent feeding liberates virus particles into healthy plants. While aphid feeding may distort young leaf growth and honeydew may foster unwanted buildup of sooty molds, the real damage is due to the transmission of tristeza virus, including those causing decline in susceptible cultivars and other strains responsible for stem pitting symptoms on citrus. Aphid species currently found on Florida citrus are capable of transmitting tristeza at relatively low levels of effectiveness, but brown citrus aphid is highly efficient in transmitting the virus, and thus poses a threat of increasing the spread of existing strains of the virus as well as other serious strains which at present are not reported to be present in Florida citrus.

Figure 34. Close up of brown citrus aphid population on citrus terminal, Dominican Republic (top); and shoot with leaf and stem infestation (bottom).

SHARPSHOOTERS, CICADELLID SPECIES

The sharpshooters, a group of leafhoppers in the family Cicadellidae (Order Homoptera), are recognized by their relatively large size for leafhoppers, their production and expulsion of large amounts of honeydew, and their role in transmission of some plant diseases. At least 3 species of this group feed upon Florida citrus. Seasonally sharpshooter populations may increase through movement between host plants to the point where they are quite visible on citrus. Young plants are particularly attractive to solitary adult sharpshooters which can be found at rest clinging to green twigs. The common species share life cycle features, including relatively long adult lifespan (up to 65 days), several generations per year, oviposition in lower leaf surface epidermis, and overwintering of adults in woody habitats. Females are capable of laying large numbers of eggs, more than 1,000 in some species. Feeding is primarily in xylem tissue, where nutrients are extracted with piercing sucking mouthparts. Young nymphs feed in leaf midveins and petioles, while older nymphs and adults utilize green stems for feeding sites.

Among species on citrus is *Homalodisca coagulata* (Say), a large species more than 11-14 mm in length. The dorsal surface is brown to black with flecks of ivory (Fig. 35). This species has been reported from some 73 plant species in 35 families, indicating the broad range of hosts on which it can develop. *Homalodisca insolita* Stal, a closely related sharpshooter species, is slightly smaller, measuring 10-11 mm in length. *H. insolita* also is dark in color with a dark brown anterior and brown wings. Close examination of morphological characteristics may be necessary to separate *H. insolita* from *H. coagulata*.

Finally, *Oncometopia nigricans* (Walker) is a large species 10-12 mm in length. The head and thorax is dark brown to black, but elytra are bluish in color with dark brown or black veins (Fig. 35). Some lighter areas of yellow appear on certain specimens of *O. nigricans*.

None of these species currently cause economic damage to citrus, and often go unnoticed within groves. However, these same species have been implicated in the transmission of such diseases as Pierce's Disease and phony peach on other hosts, and it is possible they may be capable of transmission of related diseases in citrus.

Figure 35. Adult *Oncometopia nigricans* on green stem (top); and adults (left to right) of *Homalodisca coagulata, H. insolita* and *Oncometopia nigricans* for comparison (bottom).

PLANT BUGS, *Leptoglossus* spp.

The leaf-footed bug and the citron bug (Hemiptera: Coreidae), are two species of *Leptoglossus* which attack Florida citrus. Like the southern green stinkbug, *Nezara viridula* (L.) (Fig. 37), these true bugs move as adults from non-citrus host plants on which the immature bugs develop to citrus groves in the fall as citrus fruit are maturing. Adults pierce the rind and feed on juice vesicle contents, leaving a wound through which decay organisms can enter and fruit moisture can escape. Populations of the 2 species are highly sporadic and tend to be localized, so that damage is restricted to certain groves or portions thereof. High populations can result in fruit drop and a significant proportion of fruit may be affected over a small area. Adults are good fliers and move into citrus groves from nearby host plants.

The adult leaf-footed bug, *Leptoglossus phyllopus* (L.) is approximately 1.9 cm (3/4 inches) in length and elongate oval in shape (Fig. 36). Highly visible antennae protrude forward and are tipped in yellow in contrast to a general dark brown body color of the bug. Likewise, the legs are brown with yellowish terminal segments. Additionally, a yellow band crosses the body in the region of the hind leg bases. The hind legs are enlarged in the tibial region to a greater extent than are those of the closely related and similar-appearing citron bug, *L. gonagra* (Fabricius). The citron bug reproduces on wild cucurbits such as the citron melon and on watermelon, and adults fly into groves to feed on maturing fruit in late summer and fall as melon vines mature and senesce. The adult citron bug is similar in size and shape to the leaf-footed bug, but lacks the yellow band across the body and the enlarged tibia on the hind legs (Fig. 36). Body color ranges from dark brown to black in some specimens, and yellow coloration is most noticeable along the front margin of the body. Populations of these species progress through several generations per year on non-citrus hosts, with their presence on citrus most noticeable in fall and winter. Numerous individuals may occur on a single tree and in some instances several leaf-footed bugs may be observed on a single fruit. Removal of citron vines from within groves may result in lower population pressure by the citron bug on maturing fruit.

Figure 36. Adult leaf-footed bug, *Leptoglossus phyllopus* on citrus leaf (top). Note expanded hind leg segment and yellow band across abdomen. Adult citron bug, *L. gonagra* on citrus leaves (bottom).

SOUTHERN GREEN STINKBUG, *Nezara viridula* (L.)

This small shield-shaped insect attacks an extremely broad range of host plants across the southeastern United Sates, and are often seen flying between host plants. Piercing-sucking mouthparts are used by the stinkbug to penetrate plant tissues and extract fluids, and this is the means by which damage to developing fruit occurs.

Southern green stinkbug (Hemiptera: Pentatomidae), does not normally develop on citrus plants, and thus invariably it is the adults which appear around fruit. Immatures are more likely to be found on cover crops and in vegetation surrounding the groves. Adults are oval, shield-shaped and bright green in color (Fig. 37). The insect appears rather flattened with darkened eyes and alternately colored antennal segments. Adults may be observed walking about on fruit surfaces, and when feeding they remain in place while the mouthparts are inserted into the rind and adjoining juice vesicles. A small darkened scar often results from this feeding in developing fruit (Fig. 37), and when the fruit are peeled, a discolored and dry area often appears in flesh below the rind injury. Under heavy feeding incidence by the stinkbug, fruit drop may occur. Since these types of damage are typical of all of the plant bugs which feed on citrus, the damage itself is difficult to tie to a specific pest, and the presence of the adult insect often is the only way to determine the specific cause. Adult populations of the stinkbug migrate into citrus from grove covers and from surrounding habitats, making diagnosis, damage prediction, and management difficult. Some indications of the propensity of a specific site to have stinkbug and other plant bug damage are heavy cover vegetation and proximity to dense natural vegetation. Grove margins toward the source of infestation tend to show more damage. The occurrence of this pest is highly seasonal and most stinkbugs are observed during short periods in fall and winter months when fruit are mature and host plants on which reproduction occurs are drying. Several generations (up to 4 or 5) are completed on these non-citrus host plants during the warmer months, and large numbers of stinkbugs may occur in isolated areas at one time.

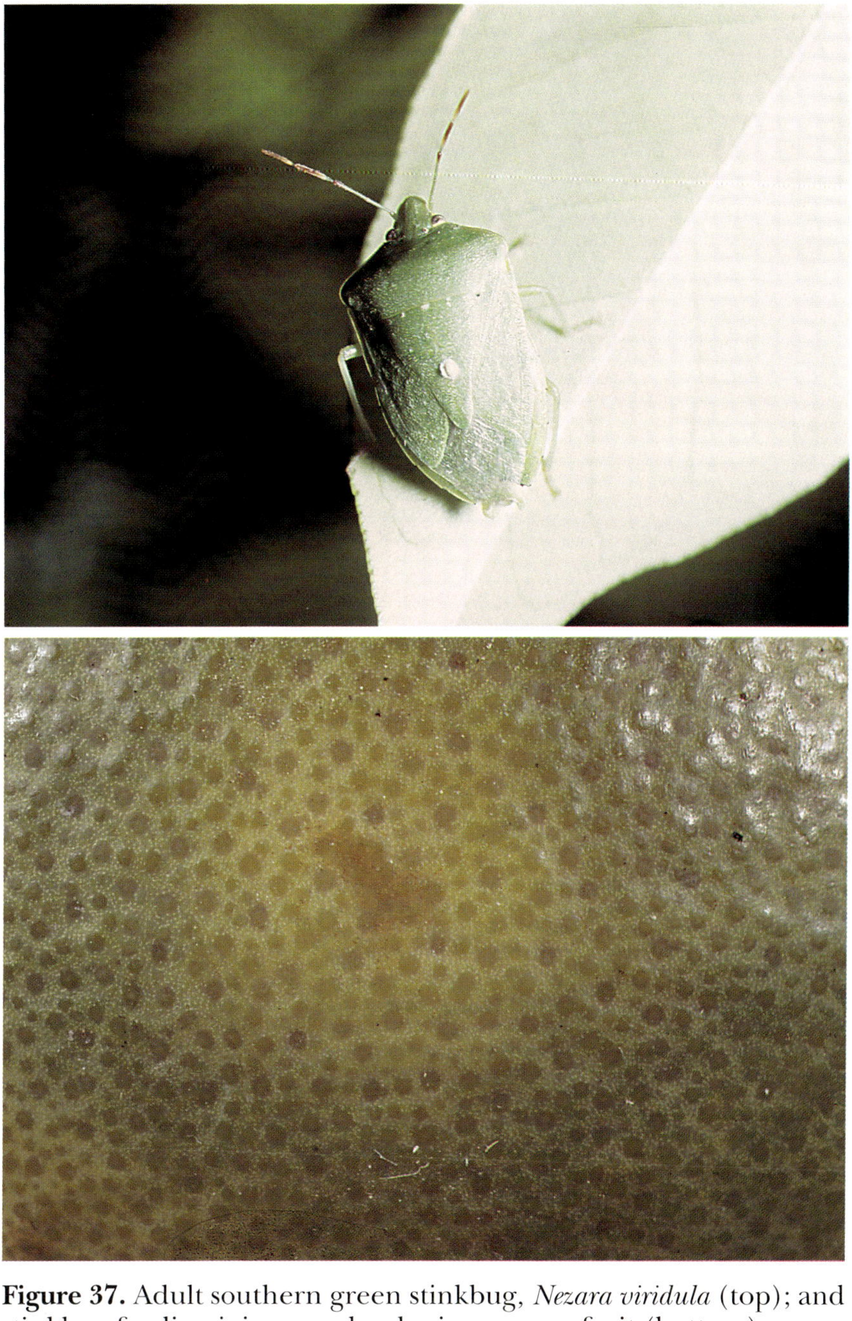

Figure 37. Adult southern green stinkbug, *Nezara viridula* (top); and stinkbug feeding injury on developing orange fruit (bottom).

CITRUS LEAFMINER, *Phyllocnistis citrella* Stainton

The citrus leafminer (Lepidoptera: Gracillariidae), is one of the most recent invaders of Florida citrus, first appearing in south Florida in spring of 1993 and representing a first record of citrus leafminer in the new world. The insect is native to Asia, where it is a serious pest of citrus. The adult citrus leafminer moth is small, measuring less than 5 mm in length. The fringed wings fold over the body and appear light in color at rest. Adults are most often found during daylight hours on undersides of leaves in the interior canopy of host plants. Tiny translucent eggs (Fig. 38) are laid singly on leaf surfaces of newly emergent citrus leaves, with feather growth 1.3 to 3.8 cm in length being preferred. Young twigs are occasionally infested, but fruit are not often attacked by this insect.

Female leafminers may lay as many as 50 eggs during a lifetime. A majority of eggs are laid on small leaf undersides, often oriented in proximity to veins. The eggs are sufficiently small and transmit leaf color so they generally go unnoticed (Fig. 38). Hatching larvae enter the leaf tissue and begin feeding on the surface layer of cells (epidermis). As development progresses, the larvae create a meandering mine in the leaf (Fig. 38). Mines normally remain restricted to leaf surface cells and do not cross through the spongy internal cells. Leafminers molt through 3 instars, and as the mine enlarges, a dark line representing larval excrement becomes noticeable. Pupation of citrus leafminer begins with the mature larva (Fig. 38) mining to the leaf edge or to a leaf fold and creating a pupal chamber while remaining inside of the leaf tissue. The pupal chamber is visible as a curled leaf edge. Transformation to the pupa occurs within the chamber (Fig. 38) and ultimately the adult emerges.

Citrus is the primary host for citrus leafminer. Citrus leafminer feeding results in citrus plants with distorted and reduced terminals. Mined leaves are reduced in photosynthetic capacity and often, full leaf expansion is not completed. Several mines may be found per leaf under high pressures, and when populations are dense, young triangular stems on emerging terminals also can be attacked by the leafminer. A generation of leafminer is completed in as few as 15 days under ideal conditions, allowing for numerous overlapping generations per year.

Figure 38. Citrus leafminer, *Phyllocnistis citrella.* Adult (top left); egg on leaf surface (left center); mature larva in exposed mine (top right); orange leaf showing two complete mines of citrus leafminer (bottom left); and citrus leafminer pupa removed from chamber along leaf edge (bottom right).

ORANGE-DOG, *Papilio cresphontes* (Cramer)

The orange-dog (Lepidoptera: Papilionidae), is a native swallowtail butterfly which ranges through much of the southeastern United Sates, with populations venturing northward during summer months. Adults are primarily black in color with yellow spots arranged in 2 rows along the margin of the wings and from the wing tip toward the central body region (Fig. 39). Adults are quite large (12.5 cm wingspan) and thus very visible around landscapes, gardens and citrus groves in Florida. Citrus plants, and especially tender young leaves, are attractive egg laying sites for the orange-dog female, where she will lay up to 500 eggs. The large eggs (about 3 mm) are round and smooth with a dull orange-cream cast, and may be easily seen on small leaves of infested trees (Fig. 39). Larvae emerge from these eggs after about 1 week and are dark brown-orange in color and approximately 5 mm in length (Fig. 39). These larvae feed openly on young leaves, consuming leaf margins and causing visible notching. As the larvae molt to ensuing instars, they grow to a length of nearly 6 cm (about 2.5 inches) and are capable of consuming enormous amounts of foliage. Mature larvae are darker in color than young larvae, being gray and irregularly marked or splotched with lighter coloration (Fig. 39). An additional feature of the larva is the ability to push out a orange forked structure concealed just behind the head capsule at rest. This structure, called the osmetrium, is visually daunting and exudes a pungent aroma which in combination with its appearance is said to deter predators.

Mature larvae pupate on citrus limbs or twigs. The pupa is dull green with yellowish markings and approximately 5 cm (2 inches) in length. As fall approaches orange-dog pupae may enter a resting stage in which they pass the winter to emerge as adult butterflies the following spring. Damage by orange-dog is restricted to leaf feeding by larvae. Young trees are particularly vulnerable to attack, since leaf damage represents a larger percentage of total leaf area. Several larvae on one plant can dramatically reduce the canopy, thus limiting growth. Some natural enemies are known to attack orange-dog larvae and pupae and thus reduce the damage of this insect under grove conditions.

Figure 39. Orange-dog, *Papilio cresphontes.* Orange-dog egg on young terminal leaf (top left); 2nd instar larva and feeding damage (top right); mature orange-dog larva (bottom left, Courtesy R. L. Phillips); and adult orange-dog butterfly on *Lantana camara* (bottom right).

SUGARCANE ROOTSTALK BORER WEEVIL, *Diaprepes abbreviatus* (L.)

The largest of the Florida weevils attacking citrus, *Diaprepes abbreviatus* (Coleoptera: Curculionidae) became permanently established in Florida after being detected in the Apopka area in 1964. Currently the weevil is commonly referred to as *Diaprepes.* The weevil has spread in the state, where it is becoming increasingly visible as a serious threat to the industry through infestation and tree death in a number of scattered locations, including southwestern Florida.

Adult *Diaprepes* are 1 to 2 cm in length and display an array of colors. Bands and splotches of orange, yellow, and white run longitudinally along the wing covers (elytra), covering the black background of the insect (Fig. 40). The snout is extended forward of the eyes and bears a pair of black clubbed antennae toward the end. Adults are slow moving and display nocturnal or crepuscular activity patterns, feeding, mating and laying eggs during late evening or early morning hours. When disturbed, adults of this and the other root weevil species drop from their resting or feeding positions. Adult feeding on leaf margins of newly emerged foliage indicates infestation of root weevils (Fig. 40), and the size and orientation of the leaf notching is characteristic for certain species. Mated female *Diaprepes* lay eggs in concealed sites, most often in the space between 2 adjacent leaves. Egg masses are deposited in a gelatinous cement which seals the leaves together and thus provides protection for the eggs (Fig. 40). Hatching larvae (Fig. 40) subsequently fall from leaf sites to the soil, where they begin feeding on small diameter roots. As feeding and growth progress, the white to cream-colored weevil larvae molt and their size increases, as does the diameter of the roots fed upon. Mature larvae attain a length approaching 2 cm (0.8 inch.) and can be found attacking root tissue in the crown area of the tree. Pupation occurs in the soil, and emerging adults dig their way to the soil surface. Two generations per year occur in Florida, each generation requiring several months. Adult longevity is several months for females and one month or more for males. Above-ground symptoms of feeding by *Diaprepes* are foliar wilt, canopy thinning, slowed growth and reduced productivity.

Figure 40. Adult *Diaprepes abbreviatus* weevil (top left); adult feeding injury to young flush (top right); *Diaprepes* egg mass on leaf with top leaf pulled apart to expose eggs (bottom left); and newly hatched larvae ready to drop to soil (bottom right).

LITTLE LEAF NOTCHER, *Artipas floridanus* (Horn)

One of the 5 root weevils attacking citrus in Florida, little leaf notcher (Coleoptera: Curculionidae), is a native species. This species is the smallest of the 5, and is not generally considered to be a serious pest of citrus. In addition to citrus, *Artipas* attacks shrubs and other trees, including native and introduced plant species. Adults of this species rarely exceed 0.6 cm in length, and are grey to dirty white in color (Fig. 41). Unlike *Diaprepes abbreviatus* (L.), there is no distinctive banding pattern or contrasting coloration on adults of little leaf notcher. Like other root weevils species, however, (see discussion of *Diaprepes abbreviatus,* page 82) adults of little leaf notcher are leaf feeders, and reproductive females lay eggs between leaf surfaces, forming a mass surrounded by gelatinous cement. Eggs are white when freshly deposited, turning more yellow as development progresses.

Hatching larvae fall to the soil and enter to begin feeding on roots. Larvae of this species are white, c-shaped, scaled in size to adult proportions, and thus remain on smaller root diameters during feeding. A chamber within the soil houses the pupa, where metamorphosis proceeds and after some 20 days the adult weevil emerges. A total of 2 to 4 months is required to complete a generation, and 3 or more generations per year are completed. Populations grow during the warmer months of the year in infested groves, and localized hot spots develop. The flightlessness of this species limits within and between grove movement to walking or passive transport on plant material. Thus, this species is not reported to spread rapidly within seasons.

Figure 41. *Artipas floridanus* leaf notching on ornamental citrus relative (left); and adult weevil on stem (right).

CITRUS ROOT WEEVIL, *Pachnaeus litus* (Germar)

The citrus root weevil (Coleoptera: Curculionidae), is one of two closely related species of curculionidae which can be found on Florida citrus, and this species is more adapted to the southern climatic zones of the state than its congener, *Pachnaeus opalus* (Olivier), the northern citrus root weevil. Populations of these two endemic species overlap in central portions of the state, where occasionally mixed populations may be encountered in the same site.

Adults of citrus root weevil (Fig. 42) are very similar in appearance and behavior to the northern citrus root weevil (Fig. 43). Blue-green in color, the broadly oval body of the citrus root weevil tapers to a broad short snout sporting grey clubbed antennae typical of the family. The hardened front wings are modified into the elytra, which are held at rest over the abdomen to protect the hind wings used for flying. The front margin of the elytra is curved with a central indentation where the two elytra meet and indentations or notches towards the sides of the body on each side. In contrast, the anterior margin of the northern citrus root weevil is smoothly rounded with only a single indentation at the elytral junction. Lines of punctations are visible running down the elytra in both species.

Adult citrus root weevils feed on succulent citrus foliage, preferring to move about and feed during hours of darkness. Like other root weevil species on Florida citrus, citrus root weevil adults remain hidden in folded or curled leaf tissue during daylight hours. Irregular notching in leaf margins indicate the presence and feeding of adult root weevils.

Mated and fed adult weevils oviposit in citrus foliage by cementing two adjacent leaves together with a gelatinous mass containing 25 to 50 creme-colored cylindrical eggs. Egg development requires a week or more, after which the white legless first instar larvae eclose from the eggs and drop to the soil. It is larval root feeding which can induce stress conditions in citrus trees which are heavily infested. Larval development ends with the mature white larva pupating in the soil. Adult emergence from the soil can be expected several months after egg hatch, and emerging adults move into the trees to begin feeding and reproduction. The total development time of several months, and adult lifespan of more than one month make for extended periods of adult presence in citrus groves.

Figure 42. Adult citrus root weevil, *Pachnaeus litus* on citrus leaf. Note notched front margin of wing covers (elytra).

NORTHERN CITRUS ROOT WEEVIL, *Pachnaeus opalus* (Olivier)

The northern citrus root weevil (Coleoptera: Curculionidae), is one of two closely related blue-green weevils which attack citrus in Florida. These two species combine with three other species of curculionids to impose root damage to trees ranging from nursery production in containers to mature trees in the grove. Adults of *Pachnaeus opalus* are blue-green in color, about 1 cm in length, and are most often encountered in folded new leaves, where they feed during hours of darkness on tender leaf tissue. Adults are broadly oval with a broad, short snout bearing grey elongate antennae with a darkened terminal club. The elytra (hardened front wings) are folded along the abdomen and the anterior junction of the elytra is indicated by an indentation along the otherwise smoothly rounded front margin of the wing (Fig. 43). This smooth anterior margin distinguishes the northern citrus root weevil from its cogener, *Pachnaeus litus* (Germar), whose anterior wing margin is curved with the midline indentation and two additional indentations, one per side.

Leaf feeding by *P. opalus* results in notching along the leaf margins, damage which indicates the presence of root weevils but does not generally cause economic loss to the tree. Numerous plant species in addition to *Citrus* serve as hosts to this weevil species, and the weevil is found in the northern half of the state, overlapping in distribution with *P. litus* in Central Florida.

Adult northern citrus root weevils fly or crawl into trees from emergence sites beneath trees, and following feeding, mating and egg maturation, oviposition begins. Egg masses of 25-50 elongate, cylindrical creme-colored eggs are deposited between adjacent leaf surfaces in a thick gelatinous mass which cements the leaves together when dried. After a week or more of incubation, the eggs yield white legless larvae about 1 to 2 mm in length which drop from the leaf to the soil surface. Larvae burrow into the soil to begin feeding upon root hairs and small feeder roots. Larval development requires several months and mature larvae pupate in the soil. Generations in Florida are overlapping with peak adult populations occurring in late Spring.

Figure 43. Adult northern citrus root weevil (Courtesy C. W. McCoy). Note smooth front margin of wing covers (elytra).

FULLERS ROSE BEETLE, *Asynonychus godmani* (Crotch)

The Fullers rose beetle (Coleoptera: Curculionidae), is a widely distributed, polyphagous root weevil species which includes citrus in its extensive host range. Fullers rose beetle biology is similar to that of other root-feeding weevils on citrus with some minor distinctions. Fullers rose beetle oviposits on leaf surfaces in common with the other root weevils, but in addition, it commonly deposits small egg masses under the calyces of citrus fruit. This habit increases the importance of the pest since it is possible that eggs can be transported on fruit. Another distinction is the lack of males in Fullers rose beetle populations, leading to parthenogenic reproduction.

Adult weevils of this species have fused wing covers (elytra) and thus are flightless. Brown in color and broader in relation to body length than other citrus root weevils, this species reaches a length approaching 1 cm (Fig. 44). The snout is shorter than the snout of other species, and the overall appearance of this weevil is thus more squared than in the other species. Adults are nocturnal and prefer interior canopy and fruit. Leaf feeding and egg laying occurs during darkness and adults rest in concealed locations during daylight hours.

Eggs of Fullers rose beetle are quite different from those of the other species in addition to their location under fruit calyces. The eggs range in number up to about 100 per mass, and are somewhat flattened. These eggs are deep yellow when fresh and darken as larval hatch approaches (Fig. 44). Emerging larvae are yellow and are characteristically c-shaped. Larvae develop as root feeders in the soil, and feeding characteristics are similar to those of the other species of similar size. Lateral roots are damaged by girdling and tree decline is an outcome of sustained heavy larval densities. Pupation occurs in the soil and emergent female beetles climb onto aboveground plant material to feed and deposit eggs. Moderate generation times allow for 2 or more generations per year and extended adult longevity leads to the presence of adults in all months of the year.

Figure 44. Fullers rose beetle adult on citrus terminal (top); and exposed Fullers egg mass (bottom).

BROADWINGED KATYDID, *Microcentrum rhombifolium* (Sauss.)

Fruit rind damage and leaf feeding are symptoms associated with the broadwinged katydid (Orthoptera: Tettigoniidae), and other katydid species. Katydids are common in Florida but rarely cause extensive injury to citrus. Young trees are more at risk to katydid damage than are older trees.

Adult katydids are 5 to 7 cm long, bright green, and strongly resemble citrus foliage due to the expanded wing shape which is oriented vertically (Fig. 45). The threadlike antennae are extremely long in katydids, and exceed the length of the body. A small triangular plate on the dorsum of the insect often shows some brown coloration in the otherwise green body color. Female katydids lay eggs in the margins of host plant leaves, and in citrus, a strand of brown or gray oval eggs about 3 mm in length can be seen partially inserted into leaf edges or laid on the leaf surface (Fig. 45). These eggs are most often found in mature leaves inside the canopy. Eggs hatch into small nymphs which resemble the adults, including the lengthy antennae. Wings are absent from all nymphal stages, but buds are visible in the last nymphal instar (Fig. 45). All active stages have extremely long, slender hind legs with which they can jump when disturbed. The shape, behavior, and coloration of broadwinged katydids make them easily overlooked on citrus foliage and they are often detected by the associated foliage damage.

Foliar damage by broadwinged katydids is similar to that induced by grasshoppers, and is characterized by notching or jagged feeding edges on leaf margins. This damage may be similar to that caused by other pests, including orange-dogs, and confirmation of the source of the feeding injury may require a search for the insect itself. Populations of katydids seldom become so abundant that control is necessary, but occasional populations in nurseries or young groves are cause for concern.

Figure 45. Eggs of katydid laid along citrus leaf margin (top left); adult katydid on leaf (top right, Courtesy R. L. Phillips); immature nymph with wing pads visible (bottom left); and katydid feeding damage on new citrus growth (bottom right).

EASTERN LUBBER GRASSHOPPER, *Romalea microptera* (Beauv.)

The eastern lubber grasshopper (Orthoptera: Acrididae), is a large native grasshopper which is commonly encountered during summer months in Florida in a wide range of habitats, including native vegetation and cultivated agriculture. Citrus infestations of this pest often occur when large numbers of the flightless insects migrate from low, marshy native vegetational areas into grove margins. Young trees are particularly vulnerable to this voracious leaf feeder, and high populations of lubber immatures can result in defoliation. Adult lubbers and their immatures are the species most often observed crossing Florida's highways. This species is more restricted geographically than is the American grasshopper, *Schistocerca americana* (Drury).

Adult lubbers, like the American grasshopper, are large grasshoppers, measuring up to 5 cm (2 inches) and are quite robust. Yellow in color, these adults have extremely short wings which prevent them from flying (Fig. 46). The wing buds are visible at the front of the abdomen and are marked with black and red, as are other parts of the body.

The life cycle of eastern lubber resembles that of other grasshoppers, wherein females lay eggs in masses in sandy soil in or around patches of vegetation. Eggs require several weeks to develop, and emerging nymphs of this species are brown in color. They are short but quite robust, as are all life stages of this species. Subsequent nymphal instars are predominantly black with some yellow and red markings (Fig. 46). The highly visible appearance and flightlessness of this grasshopper is balanced by its unpalatability to predators such as birds. In areas where lubbers are numerous, several individuals may attack a single citrus plant, resulting in substantial foliage loss. Advanced nymphal instars consume tremendous amounts of plant material, whereas, when the adult stage is achieved, feeding is reduced dramatically.

Only 1 generation occurs per year, with eggs overwintering and more or less synchronized development to adults requiring several months. Thus, most damage by this species occurs in early summer months when the species is still in the nymphal stages.

Figure 46. Eastern lubber grasshopper. Nymph (top); adult (center); and feeding damage on terminal (bottom).

AMERICAN GRASSHOPPER, *Schistocerca americana* (Drury)

The American grasshopper (Orthoptera: Acrididae), is a species which sporadically occurs in large populations, and during these periods of outbreak, the insect becomes a citrus pest, quickly defoliating trees that are in the area of infestation. The inland ridge area of Florida appears to foster conditions favorable for the periodic outbreak of American grasshopper, also known as the bird grasshopper. Fortunately, these outbreaks only occur infrequently, apparently most often following a series of mild winters.

American grasshopper is a large species, measuring up to 6 cm (about 2.5 inches) (Fig. 47). Antennae are thick, erect, and light in color. The body of the grasshopper is tan-brown with bands and spots on wings and abdomen. The head is characterized by the presence of a black band beneath the eye surrounded by a white band. The hind legs are enlarged like all grasshoppers, and the upper leg segment is banded with white and black. This species is related to the desert locust, renowned for its high populations and swarming behavior in African agriculture. The American grasshopper occurs widely across the eastern United States, but does not exhibit a migratory or swarming behavior.

Female American grasshoppers lay relatively large egg masses (up to about 100 eggs) beneath the soil level in sandy soils. Nymphs generally emerge from the egg masses in spring and fall generations in Florida to begin feeding on native or cultivated vegetation. First instars are green and subsequent instars are highly variable in color, ranging from brown to green with variable irregular yellow or green markings. The insect passes through 5 or 6 instars, consuming large amounts of leaf material during the 2-month developmental period. Nymphal feeding contributes more to the damage incurred than does adult feeding. Occurrence of this species in citrus results in leaf stripping and stunted growth. Fruit rind feeding reduces fruit quality and may affect mature trees.

Figure 47. American grasshopper female (top); male and female American grasshopper on citrus flush (bottom).

ASIAN COCKROACH, *Blatella asahinai* Mizukobo

The Asian cockroach is a relatively new insect invader in Florida, having been first detected in Lakeland in 1986. By 1989, the insect was observed in 10 counties, and at present, the full extent of its distribution is unknown. This species is very similar in appearance to the German cockroach, *Blatella germanica* L., being about 1-2 cm in length and elliptical in shape (Fig. 48). These two closely related species are extremely difficult of separate morphologically. Both are pale brown with dark parallel bands along the length of the thorax. This characteristic is quite visible in both adults and nymphs. Unlike the German cockroach, the Asian cockroach readily takes to the wing, resulting in adults often found on above-ground vegetation. Another dissimilarity between the two species is the outdoor habitat of Asian cockroach versus the domestic preference shown by German cockroach.

Asian cockroaches are closely tied to moist leaf litter and thick vegetation, where nymphs and adult females with egg cases (oothecae) are most often found. Adults stray farther from this leaf litter habitat, especially during periods of low light. The hours preceding sunrise and following sunset are most active for Asian cockroach, when adults are attracted to lights and congregate around lighted areas. In this manner they constitute a nuisance pest for nighttime outdoor activities.

Of greater importance to citrus producers is the tendency of Asian cockroaches to feed upon tender citrus leaf and stem growth. Large numbers of these roaches have been observed in certain groves, leading to leaf and growing point damage due to feeding.

Two generations occur each year in Florida, with no adults being present during January or February. Immature development is slow and adult longevity is long. Eggs are produced on an average of about 35 per ootheca, which the female roach carries during incubation. Eggs hatch within about 19 days, and nymphs undergo 5-7 molts over a 7 week period. Adults live 100 and 50 days, respectively, for females and males. The pest status of this new invader is at present uncertain.

Figure 48. Asian cockroach adult and feeding damage on tender citrus leaves.

RED IMPORTED FIRE ANT, *Solenopsis invicta* Buren

The red imported fire ant has been a serious pest in the southeastern United States since its introduction here from south America. Due to aggressive colonization habits and a painful sting, this ant comprises a major nuisance to humans and livestock through its movement into agricultural and residential areas. Fire ants are small dark ants whose black abdomen is darker than the reddish-brown head and thorax. The size of individuals varies considerably within the worker caste (2-4 mm), the most commonly encountered individuals in the colony. Other castes, including the queens and winged adults, are larger than workers.

Fire ants nest very successfully in a wide range of habitats, but are particularly prevalent in well drained sandy soils of Florida's citrus growing regions. Colonies of the ants are quite large, composed of tens of thousands of individuals and in some cases, multiple queens. Worker ants care for the colony and forage for food outside of the colony, generally on the soil surface or on vegetation. Workers excavate mounds at or near feeding sites that are meters away from the main colony, and often the evidence of fire ant's presence is a raised soft mound of soil (Fig. 49). When disturbed, active mounds teem with ants, and occasionally soil moisture is high enough to cause the ants to carry their young towards the surface. Various immature stages thus may be seen within the mound (Fig. 49). Misplaced steps into such a mound result in numerous painful stings.

Fire ants can be a major nuisance to workers, who must avoid fire ant mounds while weeding, fertilizing, picking or other grove operations. In addition to the nuisance situation created by the presence of fire ants, their presence in young citrus groves can result in direct damage to trees. The ants often follow the surface irrigation tubing along tree rows and raise mounds near trees (Fig. 49). If trees are protected from cold temperatures by wraps, the ants may establish colonies behind the wraps. Feeding by these ants on the bark results in physical damage to the trunk, access for entry of soil pathogens into the trees, and in some cases, tree death. Fire ant colonies are difficult to eliminate, but treatments are used to reduce the per acre density of mounds and thus the likelihood of damage to the trees and nuisance to workers.

Figure 49. Red imported fire ant. Disturbed mound with workers carrying larvae and pupae to shelter (left); and young citrus tree with fire ant mound along irrigation tubing. Tree wraps provide protected feeding site (right).

SUBTERRANEAN TERMITE, *Reticulitermes flavipes* (Kollar)

Termites normally do not feed on living plant material, particularly agricultural crops, and thus are not usually considered pests. However, in some Florida citrus groves subterranean termites have become known as pests of young trees. Colonies of subterranean termites (Isoptera: Termitidae), are common throughout the southeastern U.S., with soil conditions and food supply affecting colony success. Dead and decaying woody plant material is their primary food source. Microbes within the digestive system of the termites assist in the digestion of cellulose, allowing the termites to utilize woody plant material effectively. Building materials and wooden structures are susceptible to attack when the wood material is accessible to the termites.

Subterranean termites are typical of many termites, being about 5-6 mm in length and milky white to yellowish in color (Fig. 50). Termites are very soft-bodied and thus require high moisture levels to prevent desiccation. They tunnel through soil to locate woody material, and excavate the wood by burrowing throughout. Over time, the wood is riddled with these tunnels.

Subterranean termites have become a pest of citrus where young citrus trees have been planted in soil containing high population densities of termites. Areas such as pine woods where termites are important components of organic material breakdown and soil building support high populations of the termites. When these woods are cleared and citrus trees are planted in short succession, high densities of termites exploit the young citrus root and trunk material. Resulting damage is caused through reduced root growth and more importantly, physical damage to the crown of the plant (Fig. 50). Crown damage often results in tree death or severe growth retardation. Whole-tree symptoms of termite damage resemble those of many other soil pests and diseases, but closer examination often yields evidence of the feeding injury or the termites themselves.

Figure 50. Damage to young citrus tree crown from termite feeding (top); and subterranean termites in soil/wood interface (bottom).

CARIBBEAN FRUIT FLY, *Anastrepha suspensa*

The Caribbean fruit fly (Diptera: Tephritidae) is one of many economically important fruit flies worldwide which attack citrus crops. Found in Florida and south into the Caribbean region, this fruit fly has numerous hosts that are preferred over citrus for oviposition.

Adult *Anastrepha suspensa* are yellowish-brown with various bands of brown traversing the abdomen (Fig. 51). The wings are transparent with dark patterned areas typical of many fruit fly species. Adults are 12-15 mm in length with a wingspan of about 1.5-2 cm. Females possess a visible ovipositor which extends posterior from the tip of the abdomen.

Caribbean fruit fly eggs are laid in small groups just below the rind of citrus fruit. Oranges and grapefruit are susceptible at maturity, whereas, immature fruit are not susceptible to attack. Females puncture the rind and deposit the eggs in the membranous area between rind and fruit sections. Eggs are yellow in color, and elongate-oval. Hatching larvae are typical maggots, yellow in color, 2-3 mm in length and possess a pair of small dark mouth hooks. Larvae wriggle through the fruit segments during their developmental period, introducing bacteria which soften the fruit and aid the flies in digestion. Two molts occur within the fruit and third instar larvae leave the fruit, drop to the soil and pupate within the soil. Adult flies subsequently emerge from the puparium and push to the surface.

Multiple, overlapping generations occur each year in southern Florida. Citrus does not play a significant role in determining seasonal populations, as there are numerous other host plants on which the fly develops in south Florida. The Caribbean fruit fly imposes the need for regulatory control in areas where fruit destined for export is produced. A federal program involving bait spraying and sterile release methods is aimed at creating fly-free areas.

Figure 51. Adult female Caribbean fruit fly, *Anastropha supsensa* (Courtesy J. W. Lotz).

MEDITERRANEAN FRUIT FLY, *Ceratitis capitata* (Wied.)

Ceratitis capitata, commonly referred to as the Medfly (Diptera: Tephritidae), is a fruit fly of unsurpassed status. With hundreds of commercially important host plants, this fruit fly attacks citrus through much of the subtropics and is a citrus pest in many regions. At present, the fly is not known to be permanently established in North America. Infestations in Florida during 1929-30 and 1956 were eradicated using host fruit removal, bait sprays, and sterile male releases.

Medfly is smaller than the common housefly, and sports a yellowish body with white, brown and black accents. Two white bands on the abdomen and black spots visible on the thorax are characteristic (Fig. 52). The abdomen of Medfly is more rounded than that of many other fruit flies, particularly those in the genus *Anastrepha*, including Caribbean fruit fly.

The biology of Medfly on citrus is similar to that described for *Anastrepha suspensa*, except that Medfly is much more serious in its attack. Two to 30 eggs are laid beneath fruit rind in a group, and hatching larvae feed within the fleshy segments. Larvae leave the fruit to pupate in the soil, and fruit infested by Medfly are susceptible to dropping. Fruit are susceptible at maturity, as indicated by color break, but immature fruit are immune from attack. Likewise, lemons and sour lime are immune to attack by Medfly.

Possessing a relatively short generation time, Medfly can complete multiple generations per year, utilizing available fruit species in succession. Most deciduous and subtropical fruit are vulnerable, with fleshy fruit tissue, husks, or seeds providing the substrate for larval development. Up to 800 eggs may be produced by a female over her life. Fruit quality loss, fruit drop and related reduction in fruit production are symptoms of Medfly infestation. This fly is subject to quarantine barriers where it infests fruit grown for fresh marketing.

Figure 52. Adult male Mediterranean fruit fly, *Ceratitus capitata* (Courtesy M. C. Zapater).

ARMORED SCALE PARASITES, *Aphytis* spp.

Aphytis (Hymenoptera: Aphelinidae) is a genus of small wasps which are important natural enemies of armored scale insects. These parasites attack scale insects and lay their eggs upon the body of the scale. The adult wasp is minute, about 1 mm in length, and generally is yellowish in color (Fig. 53). Two pairs of transparent wings are held flat over the body, and short clubbed antennae are held forward while the active adults run about on leaf and fruit surfaces. Species are very similar in appearance, and specialists use subtle differences in wing and body structure to separate species. *Aphytis* eggs are typical hymenopteran eggs, being oblong, transparent with rounded ends. The eggs are laid either on the dorsal surface of the body by drilling through the armor cover, or are laid beneath the scale body adjacent to the leaf or fruit surface. Upon hatching, the larva is light yellow and maggot-like in appearance. Larval feeding continues for about 1 week, at which time the mature larva is approximately 1 mm in length (Fig. 53). The dead host scale is shrunken and darkened. The mature larva molts to the pupa after voiding the gut contents in the vicinity of the pupation site. The resultant dark pellets are known as "meconia" (Fig. 53), and are characteristic for this genus of parasites. Pupation occurs on the host remains beneath the scale cover, and after about 1 week, the new adult parasite emerges from the scale cover by chewing an elliptical hole in the armor (Fig. 53).

Aphytis generally develops as a solitary parasite, but in some cases, up to four or five parasites can develop on a single host (Fig. 53). Most *Aphytis* are host-specific, whereas, a few species are known to attack a range of armored scale insects across several genera. Adult female *Aphytis* can lay 2 or 3 eggs per day. Many species of *Aphytis* play prominent roles in biological control of pest scales on Florida citrus, with several having been intentionally introduced for the purpose of reducing host scale populations.

Figure 53. *Aphytis,* parasites of armored scale insects. Two *Aphytis* larvae feeding on female Florida red scale carcass (top left); three pupae and remains of Florida red scale host (top right); adult *Aphytis holoxanthus* parasite (bottom left); and Florida red scale armor showing emergence holes of *Aphytis holoxanthus* parasites (bottom right).

Encarsia PARASITES, *Encarsia* spp.

Encarsia is a genus of Aphelinidae containing many important species of biological control agents found on pests of Florida citrus. These minute wasps are parasites of scale insects and whitefly, and several have been instrumental in management of key citrus pests.

The biology of this group is somewhat varied, illustrating some of the diversity found in parasitic hymenoptera. Adult wasps are minute (1-2 mm) and generally light in color, being yellow to tan. Some species have bands or areas of darker coloration (Fig. 54) which help to visually separate them from other species. While certain *Encarsia* species have conventional reproductive strategies with males and females mating to produce fertile eggs, other species are uniparental. The population is composed entirely of females and thus they reproduce parthenogenically. Eggs are laid inside host insect bodies through the short needle-like ovipositor, after which hatching maggot-like larvae feed and develop within the host body. In transparent hosts such as whiteflies, the larva can be seen through the host exoskeleton (Fig. 54). After completing the larval cycle and killing the host, *Encarsia* larvae then pupate within the host, characteristically producing waste pellets which discolor portions of the host internally (Fig 54). Adult emergence subsequently occurs by adults chewing a hole in the host exoskeleton.

Encarsia usually develop solitarily, one parasite per host. However, another interesting aspect of their reproductive biology is that in some species, male parasites develop by feeding on other parasites. This form of hyperparasitism results from female *Encarsia* ovipositing male eggs on host scale or whitefly that already contain a parasite larva of their own or another parasite species. In this biparental system, females are produced from fertile eggs and males are produced from unfertilized eggs. Female parasites have the ability to control egg fertilization, and thus can control the sex of their offspring depending on host condition.

Among important species of *Encarsia* responsible for biological control in Florida are *E. opulenta* on citrus blackfly, *E. lahorensis* on citrus whitefly, *E. variegata* on nesting whitefly, *E. citrina* on various armored scales, and *E. lounsburyi* also on several armored scale species.

Figure 54. *Encarsia* parasites of whiteflies. Adult *Encarsia opulenta*, parasite of citrus blackfly (top); nymph of citrus whitefly showing larva of *Encarsia lahorensis* (bottom left); and nymph of citrus whitefly containing the pupa of *Encarsia lahorensis* (bottom right).

LADYBIRD BEETLES, COCCINELLID SPECIES

Among natural enemies important to biological control of insect pests on citrus are the coccinellids or ladybird beetles (Coleoptera: Coccinellidae). These beetles are predacious on a wide range of host insects, and in Florida citrus, different species can be found attacking aphids, armored scale insects, soft scale insects, mealybugs, whiteflies, leafminers, and mites. The complex of ladybird beetles represents a combination of native species and those intentionally introduced for classical biological control of specific pest insects.

The life cycle of coccinellids is typically completed within 15 to 25 days, during which all active stages act as predators. Adults have lifespans of several months in some species. Hundreds of hosts are consumed during larval and adult feeding periods. Elliptical eggs are generally laid in the vicinity of prey populations, and range in color from beige to yellow for most species. They may be laid singly or in clusters (Fig. 55). Hatching larvae are variable in shape and size, but are often elongate and roughly textured (Fig. 55). Six pairs of legs are clearly visible in most species, while in some, secretion of wax filaments obscure the entire larva. Pupation often occurs on the plant surface near the feeding site. Adults vary widely in color and size, from small brown adults (2-3 mm) of *Delphastus* attacking whitefly to large (8-10 mm) black adults sporting a pair of red spots in *Chilocorus* attacking armored scale insects. Species important to biological control in citrus are *Cryptolaemus montrouzieri* on citrus mealybug, *Chilocorus stigma* on purple scale and Florida red scale, *Azya orbigera* on soft scales (Fig 55), and *Delphastus pusillus* on whiteflies.

Ladybird beetles are most visible when attacking aphid populations on new flush terminals, but through close observation a large selection of species of this important group of predators can be found in most citrus groves. Whereas, ladybeetles are not always rapid in their response to a building pest population, their voracious appetite allows them to consume large numbers and overcome high populations of their prey.

Figure 55. Coccinellid (ladybird beetle) predators found in citrus. Coccinellid eggs on leaf near aphid colony (top left); larvae and adult of aphid-feeding coccinellid species (top right); and scale feeding adult coccinellid, *Azya orbigera* feeding on brown soft scale (bottom).

LACEWINGS, NEUROPTERAN SPECIES

Among the important generalist predators which interact with pest arthropods in Florida citrus are the lacewings, named for the net-veined clear wings typical of adults of these families. Predators in this group belong to three families of the order Neuroptera. Most familiar are the green lacewings (Family Chrysopidae), relatively large members of the order characterized by fragile-looking green adults. A smaller version of this group can be found in the brown lacewings (Family Hemorobiidae) (Fig. 56). Finally, a group of tiny net-winged predators which feed upon mites and small insects are the dusty-wings, Family Conioperygidae.

A shared attribute of these families is that all active stages are predators, with the preferred prey species varying in size and type. Green lacewings are the most successful with larger citrus insects, capable of feeding on lepidopterans, aphids, scale insects, and the eggs of many other groups. Similarly, brown lacewings feed on a wide range of prey which are necessarily smaller and/or weaker than those available to green lacewings. Finally, Coniopterygids are restricted to minute prey such as mites and tiny insect eggs. Larval prey may vary from that of the adults, and prey size and type may vary between larval instars.

The life history of these groups also is similar. Eggs are laid singly or in loose clusters near prey populations. Green lacewings lay stalked eggs (Fig. 56) to reduce the incidence of cannibalism among hatchlings. Brown lacewings and dusty-wings lay unstalked eggs on leaf and twig surfaces in the vicinity of prey. Hatching larvae of all lacewings are voracious hunters, armed with long curved hollow mandibles used to puncture prey and withdraw fluids (Fig. 56). The larvae are elongate and resemble immature beetles, having segmented legs, noticeable body segmentation, and well-developed eyes. Larvae of some species adorn themselves with remains of prey building up a layer of protection against the elements. These larvae are referred to as trashbugs. Prey consumption within some species is high, as tens or hundreds of prey are consumed over a developmental period of larval feeding. Pupation generally takes place on the host plant near the feeding sites, and the pupal form varies by family. Most familiar is the globular, white cocoon about 3 mm in diameter in which green lacewings pupate.

Figure 56. Lacewings associated with citrus. Stalked egg of green lacewing (top left); naked larva of green lacewing (top right); and adult of brown lacewing (bottom).

PREDACEOUS MITES, PHYTOSEIID SPECIES

Florida citrus groves contain numerous species of mites. Among these are species of Phytoseiidae, a large family composed almost exclusively of predators. Phytoseiids are fast-moving, active leaf and fruit inhabitors which prey upon a range of small soft-bodied arthropods. In addition, some species consume pollen as a supplement to arthropod prey or as a major component of their diet. Free-living in all stages, phytoseiids ambush and overcome their prey after which sucking mouthparts extract body fluids from the prey.

Phytoseiid eggs are laid where prey are found, often in association with the prey themselves. Eggs may be found in spider mite webbing, in the remains of whiteflies or scale insects (Fig. 57), or along leaf midribs among free living phytophagous mite populations. Phytoseiids demonstrate a consistent aversion to full sunlight, and thus spend much of their lifetime on the undersides of leaves, their major substrate. Eggs hatch into six-legged larvae, and after some feeding, the larvae molt to successive eight-legged nymphal stages. Immature stages of most species are light in color and have soft membranous exoskeletons. Some species are tan to dark brown and have slightly harder exoskeletons. They move quickly about in search of prey. At rest, they may be found along midribs or beneath webbing. Adult phytoseiids are similar in appearance to the immatures (Fig. 57), with only increases in size and appearance of reproductive structures being diagnostic. Phytoseiids are visible to the unaided eye, but a handlens is often used to confirm the presence of these mites among the leaf fauna. Slide preparations are necessary for species identification.

Phytoseiids are important mortality factors that contribute to maintenance of mite and other pest numbers beneath injury levels. Some phytoseiid species have been utilized in classical biological control of citrus pests, and the opportunity exists for further utilization of this group of predators. However, like all arthropods, pesticide sensitivity among phytoseiid mite species must be considered when pest management is undertaken so that the phytoseiid fauna will be conserved.

Figure 57. Adult phytoseiid mites found on citrus leaves (top); and egg of phytoseiid mite attached to wax strand secreted by immature whitefly (bottom).

FUNGAL PATHOGENS OF INSECTS

There are many natural enemy organisms that contribute to mortality of pest insects and mites on citrus. Fungal pathogens are well-known as important factors in bringing about epidemics which can dramatically reduce pest populations. Various mites, whiteflies, scale insects, mealybugs, and other groups are host to fungal pathogens which increase when environmental conditions favor their development. The fungus, *Hirsutella thompsonii* infects citrus rust mite and periodically causes population crashes. Efforts have persisted over decades to expand the potential to manipulate this fungus for rust mite management.

Whitefly pests in Florida have long been associated with seasonal occurrence of "friendly fungi" which bring about population decline. The predominant species of fungus, *Aschersonia aleyrodis*, attacks and kills whiteflies in the genus *Dialeurodes* and other species to a lesser degree. A combination of high whitefly populations, warm temperature, a residual inoculum of fungal spores, and moisture lead to development of widespread disease. Whitefly immatures succumb to the infection of hyphae, and soon a mat of mycelial growth emanates from infected individuals. Sporulation leads to formation of bright orange pustules surrounded by the white halo of mycelia (Fig. 58). Disease rapidly spreads among dense foliar whitefly populations, and soon all that can be seen are laves spotted with *Aschersonia* mats (Fig. 26).

Similarly, there are fungi which develop during moist warm periods on scale insects (Fig. 58). Scale populations likewise drop to near zero after an epidemic of the fungus, and sporulation is visible as fungal growth transforms the scales into a mass of fungal material. Numerous species are involved, although not all of the specific host/fungus combinations are well defined.

Adult free living insects also are vulnerable to attack by fungi. Root weevil species are infected as they emerge from soil with fungi, including *Beauvaria bassiana*, and these adults soon succumb (Fig. 58). Symptoms of fungal infections are most easily seen when the fungus growth expands through segmental membranes to the outside of the insect cadaver.

Figure 58. Fungal diseases of arthropods on Florida citrus. Disease of cloudy-winged whitefly caused by *Aschersonia aleyrodis* (top left). Orange fruiting bodies are visible in mycelial mats on whitefly carcasses; fungal disease of green scale, *Coccus viridis* on citrus leaf (top right); and *Beauvaria bassiana* infection of adult citrus root weevil (bottom). Mycelial growth is extending from segmental margins.

CITRUS NEMATODE, *Tylenchulus semipenetrans* Cobb

The citrus nematode is an obligate parasite of citrus roots which often goes unnoticed until tree stress symptoms occur. This pest was present in Florida as early as 1913, and infests most established citrus soils in the state. Death of the tree rarely follows infestation by citrus nematode, but reduced canopy density, slower tree growth, and reduced productivity may be associated with continued root parasitization by this tiny (.4 mm) unsegmented round worm. Since above-ground symptoms are associated with root damage, no specific symptoms are characteristic of citrus nematode, and changes in tree health are dependent on variety, rootstock, soil type, and numerous other factors affecting tree health. Under certain conditions, nematode damage results in visible symptoms such as canopy thinning, whereas in other cases, subtle decline in fruit size or total yield are manifested. Even apparently healthy trees have been shown to increase in productivity following treatment for citrus nematode. Roots heavily damaged by citrus nematode promote wilt symptoms which are more pronounced under water stress conditions. Populations of the citrus nematode vary tremendously making monitoring populations difficult and time consuming.

The small size of citrus nematode requires processing of soil samples for diagnosis. Adult female citrus nematodes imbed the anterior portion of their body into root tissue and feed on cell contents beneath the root surface. Egg masses produced by these females are attached directly to the root with sticky material which also traps soil particles (Fig. 59). Hatching larvae seek feeding sites on roots. Both eggs and larvae can survive in the absence of root tissue, whereas females do not reproduce in the absence of roots.

Rootstock susceptibility varies among selections planted in Florida, with trifoliate orange showing variable levels of resistance to citrus nematode. Fine textured soils prove ideal for nematode survival and development, whereas, sandy soils like those found on the central Florida ridge are less conducive to chronic high populations of citrus nematode. Sanitation through out planting of nematode free nursery stock can be important in preventing spread of this pest to areas where it currently does not occur.

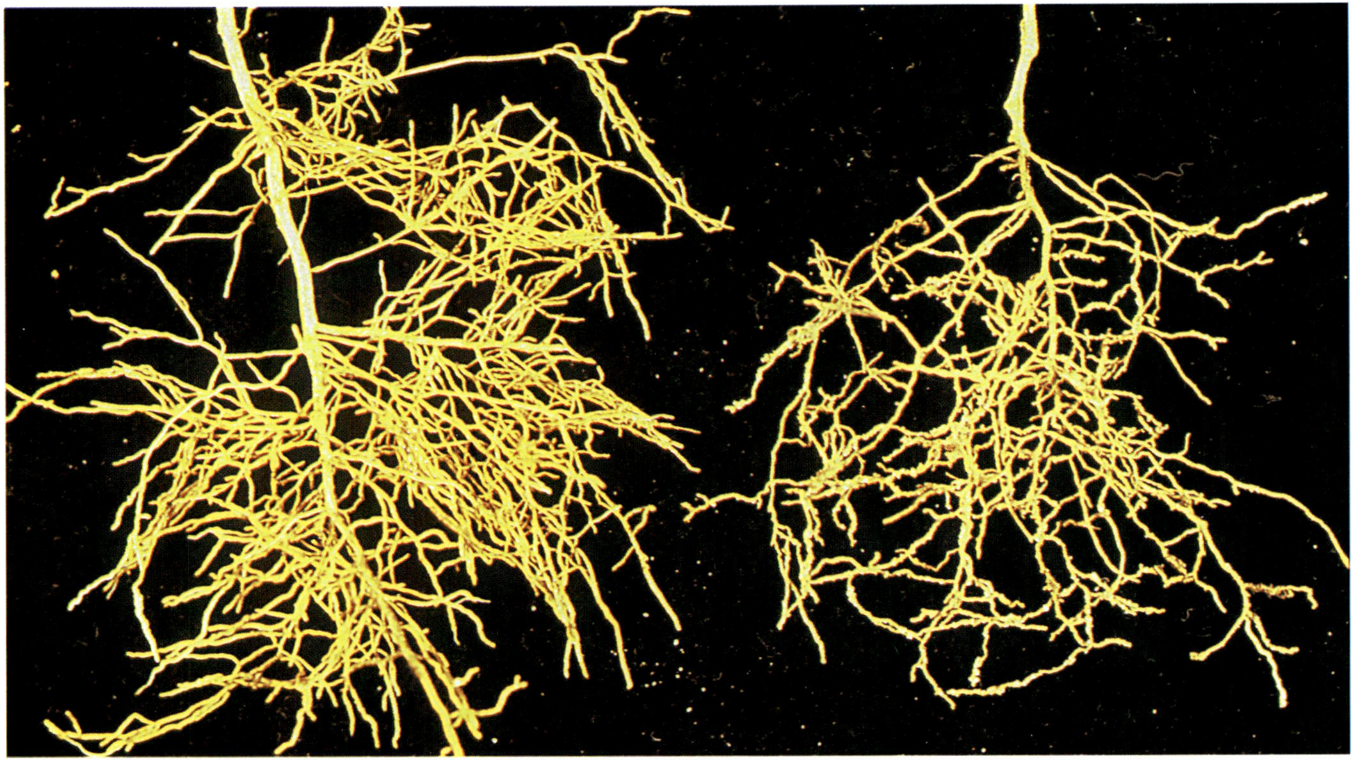

Figure 59. Comparison of healthy roots (left) with roots infected by citrus nematode (right). Note soil granules sticking to infested roots due to presence of sticky egg adhesive (Courtesy L. W. Duncan).

DECLINE PROBLEMS

Sometimes, all or part of a citrus planting fails to flourish or may even decline over time for reasons which are often not obvious. There can be many causes for such situations including virus diseases, nematodes, weevils, blight, poor soil conditions, inadequate or improper irrigation or nutrition management, poor drainage or even external environmental factors.

Whatever the cause, the symptoms are usually quite similar which makes diagnosis difficult. A general unthriftiness characterized by thinning and loss of foliage, reduction in leaf size, twig dieback and reduction in fruit size and number are the prominent symptoms. Excessive wilting and various nutritional deficiencies may also be in evidence, and in most cases, the root systems have been damaged. Consequently, nutrient and water uptake and conductivity have been impaired, resulting in canopy reduction.

Diagnosis of decline symptoms usually begins at or below ground level and possible causes are sorted out by the process of elimination. Roots should be examined for fungus, weevil larvae and nematode activity and the soil should be tested for adequate drainage. Indexing trees for virus diseases and testing water uptake for blight diagnosis may be in order. There is no substitute for extensive field experience for the diagnosis of such problems, and this should be followed up by laboratory tests for validation.

Nematode Induced Declines

Some types of nematodes may cause decline problems which are difficult to diagnose. The 2 most important problems are slow decline, caused by the citrus nematode, *Tylenchulus semipenetrans* Cobb (page 120), and spreading decline, caused by the burrowing nematode (*Radopholus citrophilus* Huettel), which is a pest of numerous crops in the tropics, but is recorded as a citrus pest only in Florida (Fig. 60). Alternate host plants allow survival and reproduction of these nematodes without citrus.

As the name spreading decline implies, this problem spreads among trees in a grove at a rate of about 2 rows of trees per year from the infection front. The burrowing nematode actually lives inside the roots, making detection difficult. Females are up to 0.7 mm long and without the need for fertilization deposit eggs within roots and in surrounding soil. Symptoms of the disorder include the same sort of general decline induced unthriftiness including wilting, nutritional deficiencies, reduced growth, yields and fruit size.

The burrowing nematode is sensitive to high soil temperatures and is rarely found near the soil surface, preferring a deeper, cooler soil location, but will be found at varying depths as conditions change. For this reason, the problem is rarely severe in bedded groves with restricted root systems although the nematode is found in all Florida citrus producing counties.

Control is best achieved through avoidance, but if the nematode becomes established, it will be almost impossible to mitigate the problem without tree removal, chemical treatment and replanting on resistant rootstocks. Careful sampling is essential to delineate the areas of infection, as often the nematode has moved beyond the area of symptom expression. Rootstocks which have shown resistance include Milam, Ridge Pineapple and Carrizo.

Water Damage Decline

Damage from excess water is difficult to describe and often involves a variety of symptoms. Chronic water damage is perhaps the easiest to diagnose and is usually easily confirmed in the field. The damage results from flooding and consequent lack of root oxygenation, followed by root injury or death, and then the demise of above-ground portions of the tree (Fig. 60). With chronic water damage, trees have very shallow root systems and are stunted with sparse foliage and small leaves. Fruit production is usually poor and fruit are small. Nitrogen deficiency (page 206) is quite common and other nutrient deficiencies often occur. When chronic water damage occurs, there are often distinct areas containing trees which have been damaged. These are often low spots, but may be areas of imperfect drainage due to proximity of a hardpan or a change in soil type. Digging inspection wells several feet deep near affected areas and observing their water levels over time will quickly confirm the potential for water damage.

Acute water damage is usually caused by some sort of flood or period of extremely heavy rainfall. In this case, the foliage will wilt, followed by partial or complete defoliation, depending upon the amount of root damage. Root condition should be checked when water damage is suspected. Injured roots will be soft or slippery and bark can be easily removed. A distinct unpleasant odor is often present due to anaerobic conditions present during flooding.

Control of the problem is by avoidance. In wet areas, provide adequate drainage with raised beds, tile drains, or avoid the area completely.

Blight

Blight, also known as young tree decline (or YTD), has been recognized as a major problem for Florida citrus for about one hundred years. Similar diseases have also been reported in Argentina, Brazil, Uruguay and South Africa. All citrus rootstocks appear to be susceptible to blight. However, rough lemon and citrange rootstocks are highly susceptible, while the onset of symptoms is delayed in sour orange and Cleopatra mandarin, and uncommon in sweet orange rootstocks.

Normally only trees which have reached bearing age are affected by blight. Initial symptoms are often only on one side of the tree and include wilting, rolling and zinc deficiency (page 212) in the leaves (Fig. 60). Spring growth is delayed and a general tree decline follows including dieback, leaf loss and reduced growth. Trunk sprouts often appear and feeder roots deteriorate as the decline continues. Abnormally high levels of zinc and stress-related compounds can be detected in the wood. Although trees never recover they seldom die, and both yield and fruit quality are greatly reduced. The bark of blighted trees remains unaffected unlike those affected by Phytophthora foot rot and psorosis. Blight symptoms include plugging and deterioration of water conducting tissue (xylem). It is more difficult to inject water into the trunks of blighted trees by means of a syringe than into healthy trees, and this is often used for field diagnosis of the disease.

Although the etiology of blight remains uncertain, the disorder can be transmitted by root grafts. The distribution of blight in a grove is at first random but later appears in groups of trees or spreads down rows. Declining trees which have lost their economic value should be replaced with those on less susceptible rootstocks.

Figure 60. Tree showing symptoms of infection by burrowing nematode (top) next to a same-aged tree with little or no decline symptoms (Courtesy L. W. Duncan). Death of trees by flooding and lack of oxygen to the roots (center). Water stress (leaf roll) and zinc deficiency in leaves of an orange tree affected by blight (bottom).

TRISTEZA

Tristeza, caused by the citrus tristeza virus (CTV), is one of the most important citrus diseases worldwide. The virus infects almost all citrus species, hybrids, and relatives and was first detected in Florida in the early 1950s. Many CTV strains exist and contribute to the diversity of symptoms expressed. Severe CTV strains have caused widespread decline and death of sweet orange, mandarin and grapefruit trees on sour orange rootstock and have almost eliminated the use of this rootstock in the state. Death of the phloem in sour orange below the bud union girdles the tree and results in overgrowth of the scion, deterioration of feeder roots, stunting, reduced fruit size, leaf yellowing, dieback, wilting and death. Tree decline may be quite rapid (quick decline) (Fig. 61) or gradual. Pinhole pitting (honeycombing) (Fig. 61) and thickening may be observed on the inner surface of sour orange bark. Rapidly declining trees may only exhibit a yellow-brown discoloration at the bud union. The most virulent CTV strains are those causing stem-pitting (deep pits in the wood under depressed areas of bark, in the trunk, branches and twigs of scions regardless of rootstock) and reductions in fruit set, size, and quality. Stem-pitting CTV strains have not been detected in Florida but are found in Argentina, Australia, Brazil, Columbia, Cuba, Venezuela, South Africa and other areas.

Spread of CTV occurs through infected budwood and by aphids. CTV is not transmitted in seed, and is not likely to move through contaminated tools in the field. CTV is transmitted in Florida by three aphid species including the melon aphid, the spirea aphid and the black citrus aphid (pages 66, 68). The brown citrus aphid (page 70), is rapidly spreading through Central America and the Caribbean and is about 20 times more effective in CTV transmission than the melon aphid, the next most efficient vector.

Quarantine against importation of CTV is important in preventing the introduction of new severe strains. Only virus-free budwood should be used, and virus-free trees planted in the field. Eradication of severe CTV strains is important to prevent and delay their distribution. Declining trees should be replaced with those on tolerant or resistant rootstocks such as Carrizo citrange, Cleopatra mandarin, rough lemon, Swingle citrumelo, and trifoliate orange.

Figure 61. Symptoms of rapid decline (left) caused by a severe strain of citrus tristeza virus. Pinhole pitting (honeycombing) in the bark of sour orange rootstock (right) infected by a severe strain of citrus tristeza virus.

GENETIC DIEBACK

Dieback of twigs, leaves and branches is one of the most common yet often inexplicable symptoms encountered among all citrus cultivars. Dieback may be caused by such noninfectious factors as cold (page 186) and lightning damage (page 194), phytotoxicity from the application of pesticides and/or fertilizers, mechanical injury, water and fertilizer imbalances which may be associated with root death, and copper deficiency (page 220). Death of shoots can be caused by algae (page 130) and is also associated with blight (page 124). In addition, a number of citrus cultivars including Robinson, Fallglo, Sunburst tangerines and Ambersweet orange appear to be genetically predisposed to dieback.

New growth dies from its tips and becomes brown and brittle (Fig. 62). Genetically predisposed dieback is usually accompanied by abundant gum exudation. It can be differentiated from other causes of dieback because it lacks the intermittent green islands produced by lightning damage (Fig. 103), or gum deposits and eruptions in twigs and corky fruit lesions associated with copper deficiency (page 124), and the zinc deficiency symptoms and leaf roll common in blight-affected trees.

The fungus *Diplodia natalensis* P. Evans which commonly colonizes dead wood and is also involved with Diplodia stem-end rot and Rio Grande gummosis, has been implicated in genetic dieback of citrus. Its role, however, may be that of a secondary invader of weakened tissue. Water imbalances and fertilizer excesses also appear to be linked to genetic dieback of citrus and should be avoided.

Figure 62. Genetic dieback in Fallglo tangerine.

ALGAL SPOT AND DIEBACK, *Cephaleuros virescens* Kunze

Although algal diseases affect many non-citrus hosts, only lemon and Tahiti lime are seriously affected citrus hosts in Florida. This disease also occurs in the Caribbean, South America, Africa, and Asia. Algal colonies appear as gray or yellowish green raised, rounded spots with feathery margins on leaves and fruit (Fig. 63). Spots on twigs and branches may coalesce to form mats which cover an area of a few inches to several feet (Fig. 63). As reproductive structures develop, algal colonies become reddish brown. Algae can cause defoliation, rind blemish, cracking of bark and branch dieback.

The causal agent is the alga *Cephaleuros virescens* Kunze (syn. = *C. mycoidea* Karst.) which survives and reproduces in colonized plant tissue. Algal spores are spread by splashing and wind blown water to new colonization sites. Algal spot is favored by the warm humid weather which occurs in summer. Management strategies for algal spot and dieback include reduction of canopy wetness through minimization of overhead irrigation, improvement of tree ventilation by pruning and weed control, and promotion of adequate drainage. Routine application of copper fungicides for control of other foliar diseases of citrus is usually effective in suppressing algal disease. However, severe outbreaks of algal spot require additional applications of copper fungicides under high pressure to wet algal mats.

Figure 63. Algal spots on lime (top, Courtesy M. Cohen). Algal colonies forming reddish mats on lemon branches (bottom, Courtesy M. Cohen).

ALTERNARIA BROWN SPOT AND DIEBACK, *Alternaria citri* Ell. & Pierce

Fruit, leaves and shoots of Dancy tangerine, Minneola tangelo and, more rarely, Honey tangerine (formerly Murcott) and Orlando tangelo are susceptible. Leaves become resistant after full expansion, and fruit 3 to 4 months after petal fall. The disease was first detected in Florida in the early 1970s and has since become a major problem in this state. Infection of young shoots and leaves causes dieback and defoliation (Fig. 64). Later infection of leaves produces discrete brown spots and/or large blotches surrounded by yellow areas. A useful diagnostic symptom of Alternaria brown spot is the blackening of leaf veins leading from dark lesions. Early infection of fruit causes slightly sunken black spots with yellow halos and fruit drop (Fig. 64). The sunken areas become raised and corky and fall out to produce characteristic round pockmarks in fruit that continue to mature. Further lesion expansion and fruit rot may or may not occur.

The fungal pathogen, *Alternaria citri* Ell. & Pierce, survives in infected shoots and produces abundant spores during wet weather. These spores readily infect lush foliage produced by heavy pruning, and excessive fertilization and irrigation. Therefore, management practices for Alternaria brown spot and dieback include minimization of shoot promoting practices such as excessive hedging and topping whenever feasible. Also avoid frequent or prolonged overhead irrigation and excessive fertilization. Multiple applications of a copper fungicide are usually required to protect fruit from infection. It is important to time sprays so as to protect newly expanding leaves and fruit.

Figure 64. Dieback and leaf spot caused by *Alternaria citri* in Minneola tangelo. Alternaria brown spot on Minneola tangelo fruit.

CACHEXIA

Cachexia has been found in most citrus growing areas of the world. Citrus hosts which are affected by cachexia, also known as xyloporosis, include mandarins, tangelos (especially Orlando), tangors, *Citrus macrophylla*, alemow and kumquats. Most other species of citrus can be infected but do not show symptoms. Symptoms of cachexia include bark scaling, stem pitting (the projections or bumps extend into the wood), discoloration and gum deposits in the bark (Fig. 65). The stem pits in cachexia-affected trees are more rounded and less pointed than those observed with tristeza. Gum deposits can be viewed by slicing through the bark with a knife, and are diagnostic for cachexia (Fig. 65). Severe symptoms include yellowing of foliage, stunting, decline and death. Cachexia is caused by a citrus viroid distinct from those which cause exocortis. The cachexia viroid is transmitted by grafting and mechanically on pruning and budding tools. It has no reported insect vector and is not seed-transmitted. Management of cachexia includes removal of infected trees and the use viroid-free budwood for propagation. Top-working infected trees with susceptible cultivars should be avoided. Budding and pruning tools should be disinfected between cuts in 0.5% sodium hypochlorite (10% household bleach).

SPHAEROPSIS KNOT, *Sphaeropsis tumefasciens* Hedges

Although all cultivars and ages of citrus trees are susceptible, Sphaeropsis knot only appears to be a significant problem on Key lime in the Caribbean area. Early symptom expression includes the formation of light colored, smooth, round knots on branches. The knots attain a size of 1 to 7.6 cm (3/8 to 3 inches) and become dark brown to black and fissured. Knots may form singly or in groups and contain numerous buds which grow out in an unrestrained fashion to produce witches brooms (Fig. 66). These knotty growths are caused by the fungus, *Sphaeropsis tumefasciens* Hedges, which produces spores in dead branches. Sphaeropsis knot appears to be rare in Florida. However, under other climatic conditions such as those occurring in Jamaica, knot formation and enlargement can girdle branches and lead to their death. Diseased branches should be pruned several inches below the lowest gall and burned.

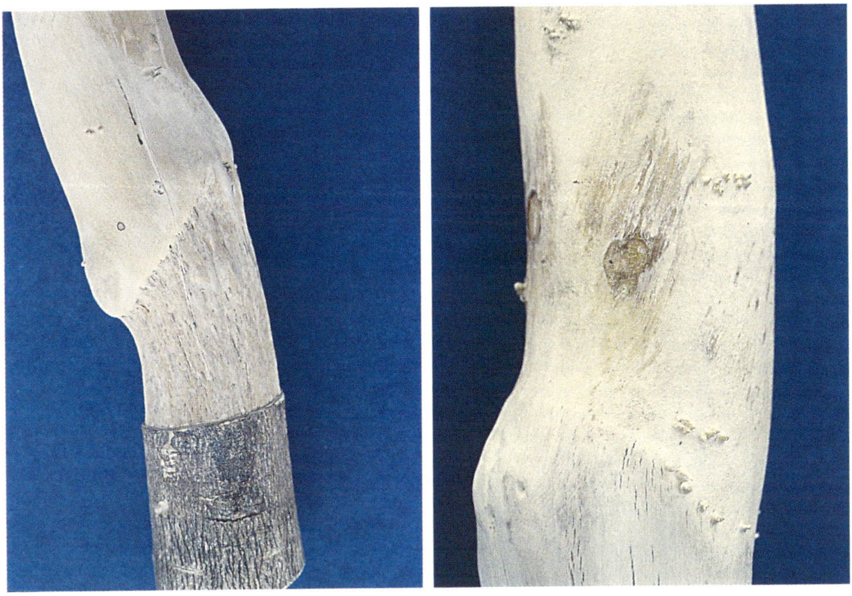

Figure 65. Stem pitting (left) caused by the cachexia viroid in Cleopatra mandarin (Courtesy R. F. Lee). Gum deposits (right) characteristic of cachexia in the wood of Orlando tangelo (Courtesy R. F. Lee).

Figure 66. Severe symptoms of Sphaeropsis knot in citrus branches (Courtesy of R. M. Sonoda).

CITRUS CANKER, *Xanthomonas campestris* pv. *citri* (Hasse) Dye

Grapefruit, Key limes, and trifoliate oranges are highly susceptible to citrus canker, whereas sour oranges, lemons and sweet oranges are moderately susceptible. Mandarins are moderately resistant, and kumquats are highly resistant. Young leaves and shoots remain susceptible for about six weeks and fruit for 3 months after petal fall. Early symptoms of the disease include pinpoint spots on leaves and fruit. As spots enlarge, they become raised, roughly circular and may be surrounded by a water-soaked area which gives rise to a yellow halo (Fig. 67). The centers of leaf spots dry, turn brown and may fall out producing a shot-hole appearance. Cankers (sunken areas) form in infected shoots and petioles and these become corky with time. Canker lesions in leaves differ from those caused by scab because they are identical on both upper and lower surfaces, are surrounded by water-soaking and yellow halos and do not produce leaf distortion. Water-soaking and the relative roundness of canker lesions in leaves also differentiate this disease from greasy spot.

The disease apparently originated in Asia and was first identified in Florida in 1913. An extensive program of eradication involving millions of trees resulted in the elimination of canker in the state by 1927. Sporadic outbreaks of the Asian strain have occurred thereafter in Florida. The bacterial pathogen of citrus canker, *Xanthomonas campestris* pv. *citri* (Hasse) Dye, survives in infected host tissue, and is spread from old lesions to new infection sites by wind blown rain, overhead irrigation and routine cultural operations. A number strains of the bacterium exist, the most severe of which has been designated the Asiatic form or Canker A. Disease development is favored by high relative humidity, abundant surface moisture and warm temperatures.

Management strategies for citrus canker include inspection and quarantine to prevent movement of infected plant material, eradication of infected trees, the use of windbreaks to reduce the spread of bacteria within groves and protective application of copper fungicides.

Figure 67. Citrus canker leaf spots in orange caused by *Xanthomonas campestris* pv. *citri* (Courtesy J. H. Graham).

CITRUS SCAB, *Elsinoe fawcettii* Bitancourt & Jenkins

Certain rootstock seedlings (rough lemon, sour orange, Rangpur lime, Carrizo citrange) and scions (lemons, grapefruit, Honey tangerine (formerly Murcott), Temple, tangelos and other tangerine hybrids) are affected by citrus scab. Sweet oranges are rarely infected in the field. Leaves are susceptible only until they are 1/4 of full size and fruit until 3 months after petal fall. Corky outgrowths begin on leaves, shoots and fruit as rounded pink pustules. Conical depressions form on the sides of leaves opposite these warty projections. Outgrowths may be so numerous on fruit that it becomes distorted and unsalable (Fig. 68). Scab can be distinguished from melanose which can also distort young leaves and shoots, by the lighter coloration and more corky nature of scab lesions.

The fungal pathogen, *Elsinoe fawcettii* Bitancourt & Jenkins, survives in old infections on shoots, leaves and fruit. Spores are produced in the scabby areas and splashed on to newly developing tissue. Temperature does not appear to be a major influence in the development of the disease. A minimum of 2 1/2 hours of leaf wetness is necessary for disease development to occur.

It is not necessary to control scab on processing fruit except in the case of Temple where severe infection may reduce fruit size. Overhead irrigation should be kept to a minimum during periods of new growth. Application of copper to decrease scab on fruit should be delayed until after petal fall because of its protective nature on the fruit. Frequent applications (every 2 to 3 weeks) of low rates of copper have been more effective in controlling scab than fewer applications at higher rates.

Figure 68. Citrus scab symptoms in the fruit and leaves of Temple.

EXOCORTIS

Exocortis is caused by the citrus exocortis viroid and other citrus viroids. It has been found in most citrus growing areas of the world and primarily affects infected citrus budded onto trifoliate orange, citranges (Carrizo, Troyer, etc.) and Rangpur lime rootstocks. The disease may be present but is latent (symptomless) in other citrus species. Symptoms of exocortis generally appear when infected trees on susceptible rootstocks are 4 to 5 years old, and include stunting and a pronounced bark scaling in longitudinal strips on the rootstock below the bud union (Fig. 69). In contrast to *Phytophthora* foot rot, the bark is not killed and symptoms do not spread to the scion above the bud union. Young trees are often stunted before bark scaling occurs. Infected trees are rarely killed by exocortis.

The citrus exocortis viroid is graft transmissible and can be spread on budding knives, pruning tools, and hedging and topping equipment, but is not seed-transmissible and has no known insect vector. A number of strains of the viroid have been identified which vary in the severity of the symptoms caused. The viroid is resistant to heat and able to persist on tools for long periods. Management of exocortis includes replacement of nonproductive trees with those on resistant rootstocks and the use of viroid-free budwood for propagation. Budding and pruning tools should be disinfected between cuts in 0.5% sodium hypochlorite (10% household bleach).

Figure 69. Bark scaling below the bud union caused by citrus exocortis viroid in Carrizo citrange.

GREASY SPOT, *Mycosphaerella citri* Whiteside

Greasy spot was first reported in Florida in 1915, and is also prevalent in other citrus growing areas with humid climates. Leaves of all commercial citrus cultivars are susceptible to some extent. The leaves of lemons, grapefruit, tangelos and early season sweet oranges are highly susceptible as is the fruit of grapefruit. Genera of plants closely related to *Citrus* such as *Aeglopsis, Fortunella, Murraya,* and *Poncirus* are also susceptible.

Yellow spots first appear on the upper leaf surface, then irregular brown blisters which become dark, slightly raised and have a greasy appearance develop on lower, and later, upper leaf surfaces (Fig. 70). Leaf drop may occur even before full leaf symptoms develop. Defoliation decreases fruit production, and makes the tree more susceptible to cold damage and attack by other pests. The fresh market quality of grapefruit may be reduced by a rind blotch consisting of necrotic specks which are at first pink and become brown or black with age (Fig. 70).

Fruiting structures of the causal fungus, *Mycosphaerella citri* Whiteside, only develop and mature in fallen leaves. When wetted, fruiting structures release spores which are carried by wind and rain to infection sites. Spores landing on plant surfaces germinate in the presence of moisture and produce tube-like hyphae which grow for some time prior to penetration through natural openings called stomates. Most stomates are found on the under surface of leaves making coverage of these areas by fungicides essential for greasy spot control. Greasy spot is favored by high relative humidity and temperature [25 to 30°C (77 to 86°F)] and by long hours of leaf wetness. In Florida these conditions generally occur during the rainy period from June through September. Late growth flushes will escape infection if there has been sufficient rainfall to aid in the breakdown of fallen leaves along with the fungal structures they contain. Symptom expression takes 3 to 4 months in grapefruit leaves and much longer in sweet orange, and 3 to 6 months for rind blotch in grapefruit.

Greasy spot management requires the application of a petroleum oil spray, or a combined oil and copper mixture for foliar infections in susceptible cultivars and for control of rind blotch on fresh market grapefruit.

Figure 70. Early foliar symptoms of greasy spot in grapefruit (top). Greasy spot rind blotch in grapefruit (bottom).

KEY LIME ANTHRACNOSE, *Colletotrichum gloeosporioides* (Penz.) Penz. & Sacc. in Penz.

Anthracnose of Key limes can result in significant losses in areas with abundant rainfall during shoot emergence, flowering and fruit development. Tahiti limes appear to be immune to the disease. Young tissue, including shoots, leaves and fruit, is susceptible. Shoots and leaves become resistant with expansion, as does fruit once it reaches about 2 cm (3/4 inch) in diameter. Symptoms of Key lime anthracnose include dieback and withering of young shoot tips, leaf distortion and necrosis (Fig. 71), and blossom blight. Infection of young fruit usually results in fruit drop. Later fruit infection produces lesions which vary from slightly sunken spots to deep cankers over much of the surface. Key lime anthracnose in fruit can be distinguished from symptoms of citrus canker because lesions lack yellow haloes, and from citrus scab because lesions are sunken rather than raised.

Key lime anthracnose is caused by a strain of the fungus, *Colletotrichum gloeosporioides* (Penz.) Penz. & Sacc. in Penz. (syn. = *Gloeosporium limetticola* Clausen) which is distinct from, but closely related to, the fungus which causes postbloom fruit drop (page 150). The fungus survives and produces spore-forming structures in diseased tissue. The irregular growth pattern and more or less continuous flowering of Key limes makes this disease difficult to control. Application of copper reportedly has brought only marginal success. Protection of developing fruit requires multiple applications of fungicide.

Figure 71. Foliar symptoms of Key lime anthracnose including distortion and necrotic spots.

MELANOSE, *Diaporthe citri* Wolf

Melanose is an important foliar and fruit disease of citrus in areas of the world where rainfall occurs during early fruit development. The disease was first recognized in Florida during the early 1890s. All commercial citrus cultivars grown in the state are susceptible but grapefruit and lemons are the most seriously affected. Leaves become resistant after full expansion and fruit remain susceptible until 3 months after petal fall.

Melanose symptoms first appear about one week after infection. The more juvenile citrus tissue is at the time of infection, the more severe the symptoms become. Early symptoms appear as small, brown sunken spots which later become raised and filled with a reddish brown gum. On leaves and green twigs, the pustules may be surrounded by yellow areas which fade with time (Fig. 72). Application of copper following lesion formation often causes the pustules to erupt in star-shaped patterns, and this symptom has been referred to as star melanose. Severely infected young shoots may become distorted and die back. Early fruit infection results in pustules spreading over large areas especially along the irregular paths (tear streaks) of water droplets (Fig. 72). A mudcake pattern results from the cracking of the margins and across the surface of infected tissue. Infection of fruit at a later stage produces smaller, round raised areas. Regardless of the plant growth stage at which infection takes place, the resultant pustules have a rough sandpaper feel, providing a diagnostic aid to differentiate melanose from the smoother discoloration caused by rust mite feeding.

The fungal pathogen, *Diaporthe citri* Wolf, develops fruiting structures and spores in recently killed twigs. Spores are blown by wind or splashed by rain on to leaves, shoots and fruit. Long periods of leaf wetness [8 to 10 hours at 25°C (77°F)] are required for disease development; the cooler the temperature, the longer leaves need to remain moist for disease to occur. Management of melanose is not usually necessary in young groves (less than 6 years old) or for processed fruit. Multiple applications of a copper fungicide may be necessary to control melanose on susceptible cultivars and in old groves with abundant dead wood. The first application of a copper fungicide should be delayed until early May.

Figure 72. Early foliar symptoms of melanose (top) showing small brown pustules surrounded by yellow haloes. Symptoms of melanose in fruit (bottom photo) includes mudcake pattern seen with early fruit infection (left) and individual round melanose pustules representative of late fruit infection (right).

PHYTOPHTHORA-INDUCED DISEASES, *Phytophthora* species

The most serious soilborne diseases of citrus are caused by *Phytophthora* species. Resistance varies by rootstock: Trifoliate orange appears to be immune to infection; Swingle citrumelo, Cleopatra mandarin, and sour orange are resistant; Troyer, Carrizo citranges, Rangpur lime, and rough lemon are tolerant; sweet orange and rough lemon are highly susceptible. Most cultivars including lemons, limes, sweet oranges, and grapefruit are highly susceptible, whereas most mandarins and their hybrids are tolerant.

Symptoms produced by *Phytophthora* in citrus include damping-off of seedlings, root rot, foot rot and gummosis. Sloughing off of outer root tissue characterizes root rot, while foot rot consists of bark peeling in crown roots and trunk at the soil level. Water soluble, reddish brown resin (gummosis) may exude from above ground infection sites (Fig. 73). Partial or complete girdling of the trunk with abundant wound tissue often occurs, leading to tree decline with yellowing of leaf veins (Fig. 73), leaf drop, dieback, yield reductions and death. Foliar symptoms may be limited to individual branches or involve the entire tree. Infection of low-lying leaves and shoots by contact with the soil are usually limited to nurseries under wet conditions. The fungal pathogens, *Phytophthora nicotianae* Van Breda de Haan (syn. = *P. parasitica* Dast.) and other *Phytophthora* species, directly enter juvenile plant tissue such as tender leaves and shoots, but need wounds and natural openings created by the root growth to invade mature tissue. A prolonged film of water on the bark from any source promotes infection.

Phytophthora management involves an integrated approach. Plant only disease-free nursery stock of resistant cultivars and avoid transplant injury. Improve the drainage of perennially wet areas before planting, and always use the correct planting depth. Time irrigation early in the morning to reduce prolonged bark wetness. Improve air circulation to promote bark drying by raising tree skirts, promptly removing wraps and soil mounds, and adequate weed control. Promptly remove and destroy new transplants with severe symptoms of *Phytophthora.* Fungicides with systemic activity may be applied to the soil, foliage or tree base. Contact or protectant fungicides such as copper must be applied preventively.

Figure 73. Symptoms of *Phytophthora* foot rot and gummosis in an orange tree (left). Yellowing of leaf veins (right) characteristic of *Phytophthora* foot rot.

POSTBLOOM FRUIT DROP, *Colletotrichum gloeosporioides* (Penz.) Penz. & Sacc. in Penz.

Postbloom fruit drop was first described in Belize in 1979, and has since been reported in Florida and Central and South America. Flowers of all citrus species are susceptible to the blossom blight phase of the disease, although the disease is more prevalent in sweet orange and Tahiti lime than in grapefruit. Navel and Valencia oranges appear to be particularly affected. All stages of flowers, are susceptible under heavy disease pressure.

Postbloom fruit drop is a rapidly developing disease and symptoms usually occur within 2 to 3 days following infection. Early symptoms consist of brown to orange, water-soaked lesions on petals (Fig. 74). These lesions rapidly enlarge, blighting individual flowers and entire flower clusters. The infected petals turn orange, become hard and dry and may remain attached to the basal disk. Fruitlets drop off at their ovary bases leaving the firmly attached basal disk and calyx (button) which continues to grow and may remain attached to the tree for a number of years (Fig. 74). When disease pressure is high, flower buds may be infected and drop without button formation.

The fungal pathogen is a strain of *Colletotrichum gloeosporioides* (Penz.) Penz. & Sacc. in Penz. that survives and produces spores on leaf surfaces and in the off-season blooms of declining trees. Infection is greatly enhanced by rain and overhead watering which readily spread fungal spores. Occurrence of the disease early in the bloom period coinciding with abundant rainfall are critical factors in causing severe outbreaks.

Management of postbloom fruit drop involves monitoring groves where the incidence of the disease has been consistently high, and fungicide application just prior to the opening of the initial major bloom of the season if the disease is present on scattered early blooms. Depending on the weather and disease pressure, multiple fungicide applications may be necessary. Prompt removal of declining trees to eliminate off-blooms and avoidance of overhead irrigation during the bloom period will also help to reduce postbloom fruit drop.

Figure 74. Early postbloom fruit drop lesions on flowers (left) caused by *Colletotrichum gloeosporioides* (Courtesy L. W. Timmer). Persistent basal disks and calyces (buttons), a characteristic late symptom of postbloom fruit drop (right).

PSOROSIS COMPLEX

Psorosis is most likely caused by a virus or complex of viruses. Most citrus species and cultivars are symptomless carriers of the psorosis virus(es). Symptomatic citrus hosts include sweet oranges, mandarins and grapefruit. Psorosis is of worldwide significance and was a major problem for citrus in Florida prior to the inauguration of a citrus budwood certification program. The disease appears to be primarily a problem in old line plantings in the state.

Symptom expression is complex and varies greatly by host with different forms of the disease; some trees remain essentially symptomless while others are severely affected. Diagnostic symptoms of a mild form of the disease, psorosis-A, include bark scaling and a dark discoloration of interior wood in the trunk and limbs (Fig. 75). Psorosis-B, a more severe form of the disease, produces mottling and ringspots in leaves and fruit, and bark lesions which rapidly expand and invade large areas. Lesion expansion may produce a girdling of limbs. The symptoms of psorosis-B are similar to those of citrus ringspot caused by the citrus ringspot virus, and the two diseases may be synonymous. Only minor amounts of gum exudation occurs with psorosis unlike the similar appearing Rio Grande gummosis. Unlike exocortis, the bark scaling observed with psorosis is not restricted to below the bud union.

Psorosis can be transmitted by budding, root grafts, and, depending on virus strain, by mechanical transmission with plant sap. Natural spread suggestive of the activity of an insect vector of a severe form of psorosis has been observed in Argentina. Management of psorosis includes replacement of nonproductive infected trees with those propagated from virus-free budwood sources.

Figure 75. Bark scaling of orange (left) characteristic of psorosis (Courtesy R. F. Lee). Wood discoloration (right) commonly seen with psorosis (Courtesy R. F. Lee).

RIO GRANDE GUMMOSIS

Rio Grande gummosis, also known as Florida gummosis, affects the bark and wood of mature grapefruit and, less commonly, oranges and tangerines. The disease is generally considered of minor importance and has been reported only on grapefruit in California and on grapefruit, oranges and tangerines in Florida and Texas.

Narrow, vertical cracks appear in the bark and yellow gum is exuded primarily in the spring and early summer (Fig. 76). Gum pockets develop and enlarge underneath the affected bark which is eventually pushed up and sloughed off in irregular scales. The exposed lesion turns a light brown and appears to heal. The process is repeated over time leading to further lesion expansion. Lesion margins are often salmon to orange colored. Wood decay follows exposure of the sapwood. Rio Grande gummosis may be distinguished from Phytophthora foot rot in that the latter causes a darker, brown wood discoloration and is generally limited to the tree base. The similar appearing psorosis lacks abundant gum exudation.

Rio Grande gummosis is a disease of uncertain etiology which has been associated with a complex of fungi including *Physalospora rhodina* (Berk. & Curt.), the sexual stage of *Diplodia natalensis* P. Evans, and *Ganoderma* species. This disease is common where breakage and death of branches have occurred especially following freezes, and branch stubs are formed by improper pruning. High chloride levels in the soil appear to exacerbate Rio Grande gummosis. Prompt removal of dead wood by using proper pruning practices can reduce the occurrence of the disease. Tools should be disinfected between cuts in 0.5% sodium hypochlorite (10% household bleach).

Figure 76. Abundant gum exudation in grapefruit characteristic of Rio Grande gummosis.

SOOTY BLOTCH AND FLYSPECK

Sooty blotch and flyspeck can affect mature fruit of most citrus species and cultivars in countries with humid climates. These diseases are usually found together and can be a problem for Florida growers after exceptionally rainy summers, especially in humid, shady parts of groves.

Sooty blotch is most often caused by the fungus, *Gloeodes pomigena* (Schw.) Colby, or occasionally by *Stomiopeltis citri* Bitancourt, and flyspeck by the fungus, *Leptothyrium pomi* (Mont. & Fr.) Sacc. These fungi generally grow superficially on the surface of citrus fruit and may rupture the waxy layer of the rind but do not penetrate the epidermis. Sooty blotch is characterized by blotches consisting of dark masses of thick-walled hyphae (Fig. 77), while flyspeck symptoms result from the production of dark, fungal spore-forming structures (Fig. 77). When fungal growth becomes so abundant that blemishes are formed, fruit grade may be lowered.

Unlike sooty mold, fungi causing sooty blotch and flyspeck adhere to the rind too firmly to be removed by washing. Sooty blotch can also be differentiated from sooty mold by less abundant fungal growth and the absence of honeydew producing insects. Routine application of copper fungicides for control of foliar diseases of citrus is usually effective in suppressing sooty blotch and flyspeck.

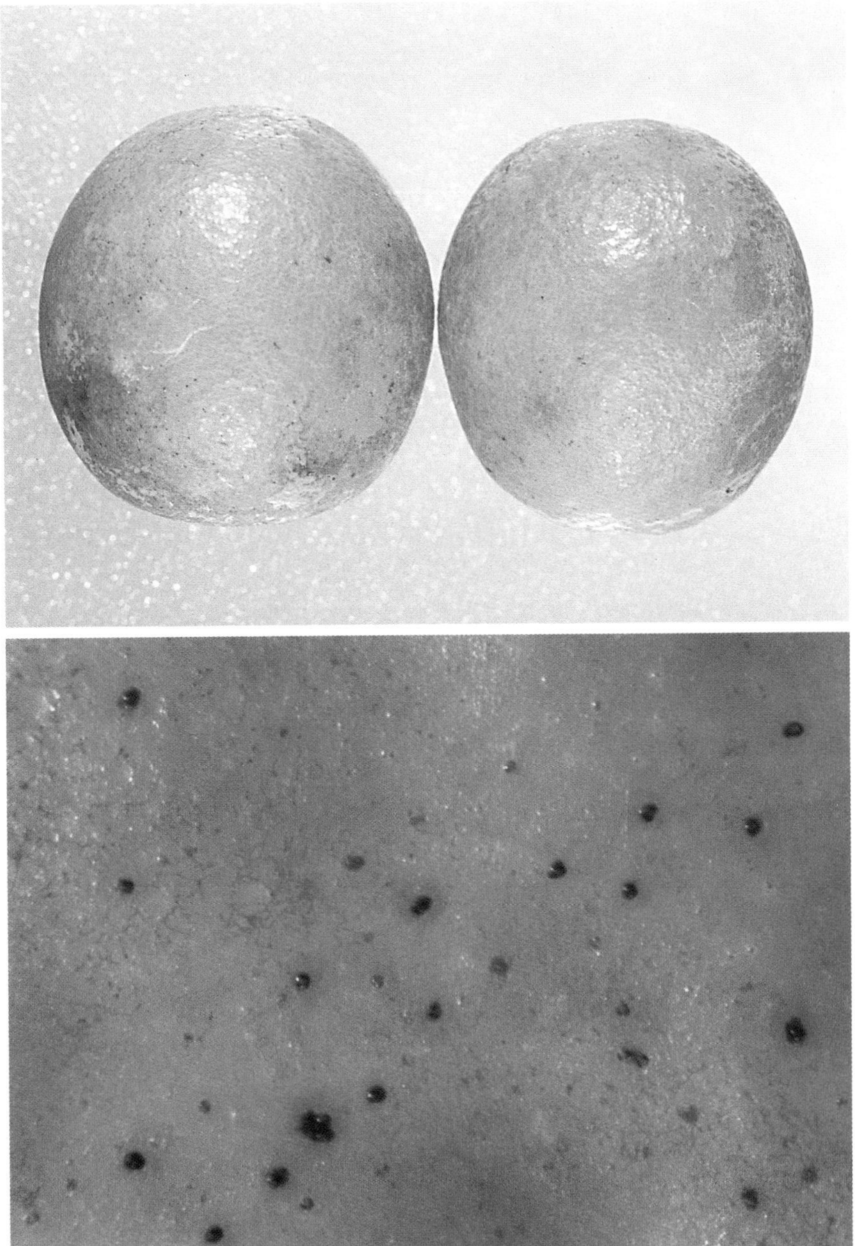

Figure 77. Sooty blotch (top) and flyspeck symptoms on oranges (bottom).

SOOTY MOLD, *Capnodium citri* Berk & Desm.

Sooty mold is the result of massive, superficial fungal growth most commonly occurring on the upper surfaces of leaves (Figs. 16 and 78), fruit (Figs. 30 and 78), twigs and small branches. Black mats of interwoven fungal strands (hyphae) may cover entire leaf surfaces and may be dislodged by the wind during dry periods. Sooty mold is usually caused by the fungus, *Capnodium citri* Berk. & Desm. which grows on the sugary excrement (honeydew) of insects with piercing-sucking mouth parts such as whiteflies, scales, aphids, and mealybugs. Although the sooty mold fungus does not penetrate plant cells, when abundant it can impair the photosynthetic functions of leaves, delay fruit coloring and present the problem of removal in the packinghouse. Sooty mold also represents a dramatic sign that insect populations may have reached damaging levels. Its control therefore involves reduction of insects causing honeydew buildup, particularly whiteflies. Application of oil, where feasible, will loosen sooty mold and may also help to control insects.

FELT, *Septobasidium pseudopedicellatum* Burt

Felt commonly occurs on the twigs, leaves and small branches of citrus trees and numerous other woody plant genera in areas of the world with warm, humid climates including Florida, the Caribbean, and parts of South America. A white, pale gray or light brown, felt-like mat encircles twigs and small branches (Fig. 79). The growth commonly covers leaf petioles and bases and radiates onto leaf blades. The growth is leathery on the surface and soft and spongy beneath.

Felt is caused by the growth (mycelium) of *Septobasidium pseudopedicellatum* Burt and other *Septobasidium* species. Certain scale insects are commonly found beneath the felt-like growth and are parasitized but not killed by the fungus. The relationship between scale and fungus is actually mutualistic in that the former gains a safe haven from unfavorable conditions and predacious insects, while the latter obtains a source of nutrients and an agent for spore dispersal. The felt fungus does not penetrate the epidermis of leaves or bark and does not appear to harm the tree. Routine application of fungicides for other citrus diseases are usually sufficient to suppress the felt fungus.

Figure 78. Sooty mold (left) on grapefruit leaf (Courtesy R. L. Phillips). Heavy sooty mold accumulation (right) on an orange and a grapefruit.

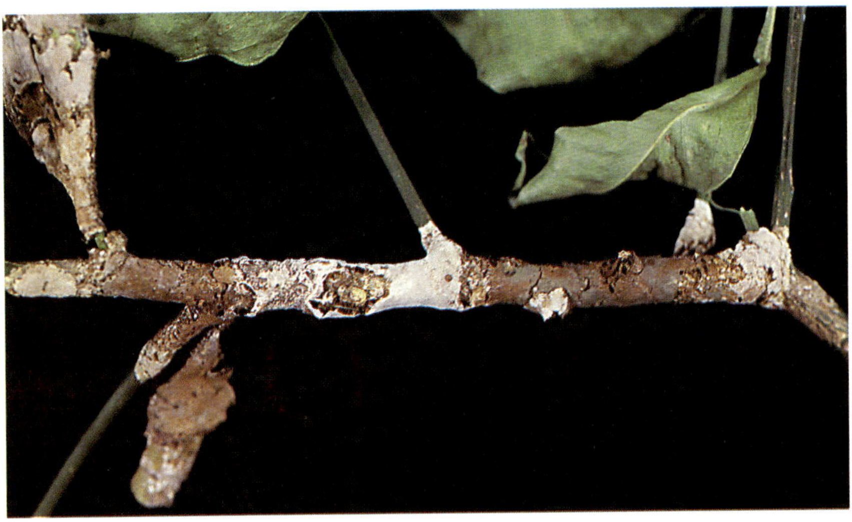

Figure 79. Felt symptoms caused by *Septobasidium pseudopedicellatum* on an orange branch (Courtesy J. W. Lotz).

WOOD ROT

Wood rot, also known as heart rot, may occur on roots, trunks and large branches of all citrus cultivars. This disease is most frequently encountered in old groves. The invaded wood is in varying stages of decay ranging from hard and discolored, to dark and mushy leading to complete internal collapse (heart rot). Trees affected by wood rot may exhibit dieback of upper branches and decline.

Many fungi can cause wood rot in citrus including *Daldinia concentrica* (Bolt. ex Fr.) Ces. & de Not., *Xylaria polymorpha* (Pers. ex Fr.) Grev., *Schizophyllum commune* Fr., *Polyporus lucidus* Leyss ex Fr. (syn. *Ganoderma sessilis* Murr.), and *Fomes applanatus* (Pers. ex Wallr.) Gill. All of these fungi produce large, easily visible spore-bearing structures (conks or mushrooms) at the bases of affected trees, and most opportunistically invade damaged and/or recently killed wood (Fig. 80). *Fomes applanatus* apparently has the ability to invade and destroy intact bark and wood adjacent to dead areas. Wounds caused by equipment, insects, hail, frost, sunscald, lightning, and pruning may act as infection sites for the invasion of wood-rotting fungi. Wood rot may also be secondary to *Phytophthora* root and foot rot. The removal of dead wood and the use of wound painting compounds have generally proved ineffective in controlling wood rot. This disease is best controlled by avoiding unnecessary wounds and by maintaining tree vigor through proper horticultural practices. It is very important that large limbs be pruned at the correct distance from the trunk to facilitate healing and avoid dead stub formation.

Figure 80. Spore-bearing structure of *Polyporus* species at the base of a tree affected by wood rot.

ANTHRACNOSE, *Colletotrichum gloeosporioides* (Penz.) Penz. & Sacc. in Penz.

Anthracnose fruit decay is caused by the fungus, *Colletotrichum gloeosporioides* (Penz.) Penz. & Sacc. in Penz. (*Glomerella cingulata*), which also occasionally causes dieback of twigs, leaf spotting and fruit tear staining. The fungus is a symptomless invader of citrus fruit, but it only commonly causes disease on degreened early season varieties such as Robinson and Fallglo tangerines, navel oranges and grapefruit. It may also develop on all cultivars at injuries caused by insects, wind, sunburn, sprays or other means.

The decay is most severe on pale, yellow colored mandarins, and least on dark orange fruit. Symptoms may appear anywhere on the fruit, but are most often at the stem end. The decay first appears as a firm silvery gray area (Fig. 81) which expands, turns a light brown, and then darkens. Degreening ethylene triggers invasion of the rind by the fungus. The fungus produces spores in dead twigs on the tree, and the spores are spread by wind and water to the fruit. Losses due to anthracnose may be reduced by cultural practices which maintain healthy trees and reduce twig death. Avoid injury to fruit and reduce degreening time and ethylene level to the necessary minimum. Delay harvest for better development of natural peel color, or spot pick for color. Degreen representative samples of susceptible crops to predict the incidence of this decay. Hold packed fruit below 4.4°C(40°F) to retard anthracnose development.

SLOUGHING

Sloughing is reported to be a physiological disorder, but it may be a severe form of anthracnose. This disorder appears only some seasons on very early red or pink grapefruit a week or longer after harvest, usually during marketing.

Sloughing occurred more frequently during the 1950's before minimum maturity standards were increased, and harvests were effectively delayed. The dark brown peel remains moist and is easily moved with finger pressure (Fig. 81). The disorder is limited to the peel and does not invade the fruit flesh. Effective control may be achieved by delaying harvest of susceptible crops. There is no known cultural or chemical control for sloughing.

Figure 81. Anthracnose (top) with distinctive silvery gray to light brown infection on Sunburst tangerines. Sloughing (bottom) involves the peel and not the flesh of red or pink grapefruit.

BLACK ROT, *Alternaria citri* Ellis & Pierce

Black rot (also known as Alternaria stem-end rot), caused by the fungus, *Alternaria citri* Ellis & Pierce, is widely distributed in Florida and other citrus-growing regions but only occasionally is prevalent enough to cause economic loss. The disease is most common on early varieties, especially navels, Jaffa and tangelos, but may occur on other oranges and tangerines. It may appear following a freeze and after long term cold storage.

The most characteristic external symptom is premature coloring on the tree. Infected fruit color sooner than normal fruit and turn a deep orange color. Many of the diseased fruit drop early in the season, while others that are infected later or are less extensively diseased may remain on the tree and appear sound on the exterior.

Infection usually begins at the stylar or blossom end of the fruit. When cut, infected fruit will be found to have a dry, black, decayed area at or near the stylar end or stem end (Fig. 82). In more advanced cases, the decay extends along the core of the fruit. Diseased fruit are likely to break down in transit and are troublesome when the fruit is processed because diseased fruit are likely to escape the attention of the graders, resulting in off-flavor, black specks and an increased mold count in the juice.

Black rot is seldom of sufficient importance to justify control measures. If infection appears in a mature crop, the crop may be harvested late in the season to allow time for the infected fruit to drop. Postharvest treatments which delay fruit button (calyx) senesence may delay the development of black rot.

Alternaria citri may infect Valencia oranges and grapefruit held in long term cold storage. This is sometimes called *Alternaria* stem-end rot, but the fungus is not related to those which cause stem-end rot described on page 172. The reason for this name is that the internal decay is often first observed at the stem-end of the fruit.

Figure 82. Black rot of Pineapple oranges sometimes exhibits external symptoms at the blossom end (top), but is more often found in the core of the fruit (bottom).

BROWN ROT, *Phytophthora citrophthora*

This fruit rot is caused by the same soilborne fungi, *Phytophthora citrophthora* (Sm. and Sm.) Leonian and less frequently by *P. parasitica* Dastur, that cause foot rot. In Florida, brown rot is not usually a serious problem but in seasons of abundant fall rain there has been considerable loss of fruit in some groves, especially on the heavier soils.

Brown rot is usually most abundant within 3 to 4 feet of the ground, but under Florida conditions, where rains are frequently accompanied by high winds, infections may be found high in the trees. The decayed area first appears 3 to 4 days after infection as a light discoloration. The infection is firm and leathery, developing to a tan in color (Fig. 83). The infected non-sunken area spreads rapidly, and may be covered by a white mycelium when the relative humidity is high. Brown rot-infected fruit have a pronounced and characteristic rancid odor. Lemons are the most susceptible, but all types of citrus may be infected.

Brown rot is less pliable than stem-end rot and is likely to occur on the side or blossom end of the fruit. The color is tan from the onset of decay, whereas stem-end rot tends to show a water-soaked area first. Brown rot may be distinguished also by the characteristic odor.

Control may be obtained by spraying the trees with copper compounds or other suitable fungicides. When brown rot is prevalent, take precautions to prevent spread of infection in the packinghouse. It is helpful to reduce irrigation near harvest. Increased ventilation by mowing, weed control and pruning limbs up from the ground provides an environment less suitable for spore production and infection. Delaying harvest of an infected crop by 2 weeks will allow diseased fruit to fall. Drops should never be picked up in a brown rot infected grove. Development and spread of brown rot is inhibited after harvest by holding fruit at 4.4°C (40°F).

Figure 83. Brown rot of Hamlin oranges with white mycelium on the surface (right).

GREEN MOLD, *Penicillium digitatum* Sacc. and BLUE MOLD, *Penicillium italicum* Wehmer

Green mold caused by the fungus, *Penicillium digitatum* Sacc., is one of the most important fruit decays in Florida. Blue mold, caused by *Penicillium italicum* Wehmer, occurs infrequently in Florida, but is more common in other climates. In Florida, molds are most common in the cool months when development of the disease is favored.

The fungus enters the fruit only through injuries to the skin. The decay first appears as watery discolored spots that are easily punctured by finger pressure. Later the white fungus mycelium appears on the surface and soon produces a mass of powdery olive-green or light to bright blue spores. As the disease advances, the colored sporulating area is surrounded by a white margin, which is wider for green mold than for blue mold. Finally the decayed fruit becomes soft, shrunken, and shriveled, and is entirely covered with spores (Fig. 84). Green mold is distinguished from other common decays in Florida by the presence of the olive-green fungus spores.

Mold sporulation may be inhibited by fungicide treatment so that the soilage of noninfected fruit with spores may be reduced or eliminated. The appearance of this fruit is a white, wrinkled firm surface (Fig. 84). In the markets this is erroneously called white mold, but is really sporulation inhibited blue or green mold. Sporulation may occur as the fungicide becomes less effective over time, thus spores appear on or at the edge of the white surface.

Losses from mold may be reduced by taking precautions to avoid injuries to the skin of the fruit. Treatment before or after harvest with approved fungicides provides control of molds. *Penicillium* can become resistant to fungicides so that sanitation in the handling, packing and storage operations is very important. Also, tests to detect resistant molds in the packinghouse need to be done periodically, especially during the cool winter months. Remove all debris and decayed fruit from the packing site and treat mold exposed equipment with an approved sanitizing agent. Repacking *Penicillium* infected fruit is best accomplished in a site remote from the packinghouse. Shipping temperatures at or below 4.4°C (40°F) will delay mold development until the fruit is warmed during marketing.

Figure 84. Green mold (top) is common in Florida fruit. Blue mold shown on Australian navel oranges (center) infrequently occurs in Florida. Fungicide inhibited sporulation of molds results in a leathery white decay (bottom).

SOUR ROT, *Geotrichum candidum*

Sour rot is caused by the fungus, *Geotrichum candidum* (*Endomyces geotrichum*), which is commonly found in Florida soils. The spores of this fungus are spread to fruit by water and wind. Fruit on or near the ground have more sour rot fungus than fruit higher in the tree. Infection is at insect or mechanical wound injury sites in the rind. Sour rot may infect all types of citrus fruit, but is most common on mandarins and very mature oranges and grapefruit.

Fruit harvested with high peel moisture early in the morning or following rain or irrigation are more likely to develop sour rot than fruit with lower peel moisture. Initial symptoms are slightly raised, water-soaked, clear to yellow lesions (Fig. 85), and are sometimes confused with those of *Penicillium* molds. When relative humidity is high, a white, yeasty layer may cover the lesion. The infection quickly spreads into a soft decaying area, and is favored by temperatures around 27°C (80°F). The name sour rot reflects a characteristic sour odor associated with this decay. Sour rot attracts fruit flies which may spread the fungus. This decay spreads from infected to healthy fruit in packed cartons.

Control of sour rot infections is difficult. Preventative measures of limiting soil contact with the fruit, limiting injuries by very careful harvesting and handling, and sanitation in field containers, the packinghouse and on the packingline are the best means of limiting losses due to this decay. Delaying harvest until the fruit moisture and turgidity lessens will help to reduce sour rot. Do not allow sour rot to contaminate the packingline by careful grading immediately after the dump. Truck drenchers and packingline soak tanks need adequate sanitation with chlorine at the proper pH, or another approved sanitizer to prevent the spread of this disease. Transportation and holding temperatures below 10°C(50°F) will suppress the development and spread of sour rot which will resume development upon warming during marketing.

Figure 85. Sour rot water-soaked soft decay on Orlando tangelo.

STEM-END ROT, *Phomopsis citri* Faw. and *Diplodia natalensis* Pole-Evans

This fruit decay is caused by the fungi, *Phomopsis citri* Faw. (*Diaporthe citri* (Faw.) Wolf), which also causes melanose, and *Diplodia natalensis* Pole-Evans (*Physalospora rhodina* (B. and C.) Cooke). The stem-end rots caused by these 2 fungi are similar in appearance and chemical control measures are also similar. Stem-end rot is one of the most destructive fruit decays in Florida. Both fungi colonize dead twigs and wood on the tree where spores are dispersed by rain and wind to fruit.

The decay usually appears as a leathery, pliable area encircling the button or stem end of the fruit. The affected area is buff colored to brown (Fig. 86). Decay caused by *Diplodia* often advances in streaks down the side of the fruit, and internally will be found to spread most rapidly in the core. *Diplodia* stem-end rot may also occasionally begin in wounds in the rind and at the stylar end of the fruit. Decay caused by *Phomopsis* spreads through the core and in a nearly even rind pattern from the stem end without finger-like projections. Stem-end rot does not usually spread from fruit to fruit in packed cartons.

Decay occurs after harvest when the fungus grows from the calyx (button) into the fruit. *Diplodia* is most serious in early season fruit which has been treated with ethylene, due to the abscission of the button that exposes natural openings for fungal pathogens. Later in the season when degreening is not necessary, stem-end rot due to *Phomopsis* becomes more common.

Cultural practices which minimize dead wood are helpful. Prune to remove dead wood, especially following a freeze, to reduce inoculum production and subsequent fruit infection by stem-end rot fungi. Harvesting by snapping (pulling) rather than by clipping will reduce infection. Degreening ethylene level and time should be kept to a minimum. Ask about currently approved predegreening fungicides which are effective on stem-end rot. Promptness in handling and shipping, and refrigeration in transit, are important in reducing losses. The disease is arrested at temperatures below 10°C (50°F).

Figure 86. Stem-end rot caused by *Diplodia natalensis* (top) usually begins at the stem end and may infect through the core to the blossom end. Note distinctive uneven progression of the infection (right). *Phomopsis citri* (bottom) stem-end rot progresses more evenly on the fruit surface.

CHILLING INJURY

Chilling injury is a physiological disorder of many tropical and subtropical crops wherein symptoms occur from exposure to above freezing cold temperatures. For citrus fruit the time of temperature exposure to initiate chilling injury is relatively long, with a minimum of approximately 1 month.

Oranges grown in other climates may exhibit chilling injury, but those grown in Florida are rarely if ever affected. Symptoms on Florida grapefruit may vary widely from the most common pitting of the peel (Fig. 87) to a mild discoloration or scalding (Fig. 87). Chilling injury pitting is more severe on the sun exposed area of sunburned fruit (page 198). Pitting of limes may also be caused by chilling injury. A collapse of discrete areas of the peel form sunken darkened lesions which will slowly expand. Scalding, also known on lemons and other types of citrus is usually a light tan color.

Chilling injury is most likely when Florida grapefruit are harvested very early and very late in the season during periods of active tree growth. They are least susceptible during the cooler winter mid-season. Postharvest low relative humidity enhances pitting making chilling injury more severe. Florida grapefruit are more likely to exhibit chilling symptoms when exposed to temperatures at or near 4.4°C (40°F) than to warmer or cooler temperatures. Grapefruit chilling injury has not been observed at or above temperatures of 15.6°C (60°F) and rarely in mid-season harvested grapefruit exposed to 10°C (50°F). Waxing and film wrapping (which reduce moisture loss) and certain fungicides are known to reduce, but not eliminate, chilling injury. Also, experimental intermittent warming during storage, although not practical for large masses of grapefruit, has been shown to reduce chilling injury.

Figure 87. Chilling injury of grapefruit with distinctive pitting (top). Scald symptoms (Courtesy R. E. McDonald) of grapefruit chilling injury (bottom).

OLEOCELLOSIS (OIL SPOTTING)

Oleocellosis, commonly called oil spotting (Fig. 88), occurs on all varieties of citrus. Navel oranges, lemons and limes are especially vulnerable to this injury. This problem is caused by oil released from broken peel oil cells in the flavedo (outer colored portion of the peel). These natural oils are toxic to the surface of the peel and cause a necrosis of the surface cells. The affected areas gradually darken and become sunken. A special form of oil spotting is caused by sand (Fig. 88) and is recognized by discrete small sunken areas in the peel. Sand damage may occur by blowing sand in a wind storm, and be found mostly on the outside of external fruit on the windward side of each tree. Sand damage after harvest will be distributed around the fruit, caused as fruit rolls in sand in the dumping or conveying process.

Oil spotting is more common when soil moisture levels are high and fruit is turgid. Early morning is usually the time of highest turgidity following the night when stomates are closed in the leaves and fruit. During daylight hours stomates open and gradually the turgidity decreases. Sand is a contributing factor to increased oil spotting by causing minute injuries during handling.

Oil spotting may be reduced by avoiding harvest each day until the fruit is less turgid. A pressure tester has proven to be valuable in estimating when turgidity has decreased enough to allow safer harvesting. A pressure tester with a standard 3/8 inch rod-like slightly rounded contact head measures the force (in pounds or kilograms) necessary to break oil cells in the peel of fruit. Moisture on a paper tissue between the head and fruit allows freshly broken oil cells to be readily detected. A pressure range of 0 to 10 pounds is adequate to detect a range of very tender peel to fruit with adequate resistance to oil spotting.

Pressure testing is especially useful on high value and more susceptible crops. Avoiding sand and other trash contact with fruit, and clean well repaired pallet bins also reduce the incidence of oil spotting. Careful and continuous supervision of harvesting is essential to avoid this problem.

Figure 88. Oil spotting of navel orange (top) and sand damage (bottom) of Orlando tangelo.

STEM-END RIND BREAKDOWN

Stem-end rind breakdown may occur on any kind of citrus, but most often occurs on oranges and Temples. The most severe and earliest symptoms are usually seen on small fruit. Thin-skinned fruit tend to be affected more than thick skinned fruit.

Symptoms usually appear after harvest and during storage at the stem end of the fruit. There is a collapse of the rind tissues resulting in a sunken brown area which is irregular in shape. The distinctive feature of stem-end rind breakdown is a narrow ring of unaffected tissue immediately around the stem area which has no stomata (Fig. 89).

This disorder appears to be associated with a nutritional imbalance involving nitrogen and potassium, particularly potassium deficiency. This relationship is poorly defined so that there are no specific nutritional suggestions except to maintain a well balanced nutritional program.

Stem-end rind breakdown is closely associated with moisture loss from the fruit after harvest and before a protective wax coating is applied in the packinghouse. Susceptible crops should be handled as quickly as possible during the harvesting to waxing interval. The fruit should be held at very high relative humidity (over 90%) during degreening or during delays before waxing. Also, excessive brushing on the packing line should be avoided as this can accelerate moisture loss and increase stem-end rind breakdown. Automatic packingline brush wipe-outs will reduce time on the brushes and may reduce the severity of stem-end rind breakdown.

Figure 89. Stem-end rind breakdown of Hamlin oranges with distinctive area of unaffected tissue immediately around the stem-end.

STYLAR-END BREAKDOWN OF LIMES

This non-pathogenic disorder occurs in seedless Tahiti or Persian limes (*Citrus latifolia* Tan). Stylar-end breakdown, has, at times, been mislabeled stylar-end rot because the symptoms have the appearance of a progressive decay. It first appears as a tan water-soaked area on harvested fruit, most commonly near the stylar end (Fig. 90), and may also develop from the stem end. It can occur within hours after harvest of otherwise normal fruit, the affected area spreads rapidly. *Penicillium* mold (see page 168) is frequently a secondary infection during shipment but rarely infects otherwise healthy limes. Stylar-end breakdown occurs in Florida during the summer months. Another summer disorder which is sometimes mistaken for stylar-end breakdown is yellow tip (Fig. 90). As the name suggests, a yellowing of the stylar-end of limes during maturation on the tree is the first symptom. Yellow tip develops a water-soaked lesion while the fruit are still attached to the tree and dries to a sunken, dark peel area.

Stylar-end breakdown occurs as a result of ruptured juice vesicles in the margins of mature, turgid, limes. This is caused by post harvest solar heating of fruit picked during periods of high turgor pressure. The liberated juice invades the rind at the stylar end and at times the stem end via the central core. The acidic juice causes breakdown of chlorophyll in the flavedo. The rate of spread and level of incidence depends upon fruit turgor pressure and level of post harvest field heat, especially in larger fruit.

Fruit having rind oil release pressures of 3.0 kg (6.6 pounds), using a pressure tester equipped with a 9.5 mm (3/8 inch) head, are generally safe for harvest as long as the fruit are not allowed to accumulate field heat. The problem is most likely to occur if fruit are picked when turgor pressure is high (early morning or soon after rains) and the fruit are temporarily stored in the summer sun prior to transport to packing facilities.

Control is realized by avoiding harvest of the largest fruit, picking late in the morning when the fruit are less turgid and keeping harvested fruit in the shade. Incidence of oil spotting (see page 176) is also associated with high turgor pressure. Careful handling is necessary to prevent rupture of rind oil glands.

Figure 90. Stylar-end breakdown (top, Courtesy M. A. Trunk), and yellow tip (bottom, Courtesy T. L. Davenport) of Persian limes.

BLOSSOM-END CLEARING OF GRAPEFRUIT

This disorder, also known as stylar-end clearing, water log and wet bottom, occurs most frequently late in the season on high quality thin-skinned grapefruit. It is found earlier and more frequently on red than on white varieties. More severe cases may lead to decay.

Ruptured juice vesicles in the core of the fruit are associated with bumps or drops. Juice in the center cavity soaks through the peel, most frequently at the sylar-end. This creates a water-soaked, translucent appearance about 24 hours after the juice vesicles are broken (Fig. 91). In very mild cases of this disorder, the translucent area may disappear by reabsorption or diffusion of the juice, but in most cases the symptoms persist.

Higher pulp temperatures and lower humidity are reported to increase, but not cause blossom-end clearing. Also, small fruit are more severely affected than large fruit. Control may be achieved by very careful handling of susceptible crops. Also, seasonally earlier harvest of thin skinned grapefruit will help to avoid this problem. Precooling before handling in the packinghouse will reduce the incidence of blossom-end clearing.

RUMPLE OF LEMONS

Rumple is a nonpathogenic disorder of fresh market lemons grown in Florida and the Mediterranean area. Lemon juice and oil are not affected by this disorder. Rumple has been reported in Sicily with the descriptive term, wrinkle rind. In Florida, rumple initially appears in late summer as chlorotic rind spots which include several oil glands. The oil glands develop a greenish color (Fig. 92) and gradually change through tan and brown to black. The name rumple is due to the wrinkled appearance of the affected peel.

Rumple occurs in varying amounts, but may affect 75% of the crop of some varieties in some years. Large fully mature fruit are most likely to develop this disorder. Also, fruit from trees receiving large amounts of nitrogen fertilizer are more likely to be affected than from trees with limited nitrogen. The only control of losses is by harvesting early and by reducing nitrogen fertilization.

Figure 91. Blossom-end clearing of grapefruit results from juice soaking through the peel, most often through the center cavity to the blossom-end of the fruit.

Figure 92. Rumple on a lemon fruit with greenish wrinkled peel.

PLUGGING

Plugging is the removal of a portion of the rind when the fruit is pulled from the stem during harvest. Fruit with large plugs (Fig. 93) may be removed by graders on the packingline. Small plugs and slight rind tears are much more difficult to see in fruit and these often develop decay during marketing.

Any type of citrus fruit may be plugged, but loose skinned mandarins are most likely to exhibit this problem. At times oranges are very difficult to separate from their stems at harvest and are therefore plugged.

Avoidance of plugging by careful harvesting practices is advisable, especially for high value mandarin crops. Clipping, when done correctly without protruding stems, is sometimes an economically viable alternative for plugging. Oranges, grapefruit and tight skinned mandarins, such as Honey tangerines, should be harvested by a snapping action rather than by pulling to reduce plugging.

ZEBRA SKIN

Zebra skin, usually of tangerines, appears as a darkening of areas of the peel over the individual fruit segments (Fig. 94). This alternate stem-to-blossom pattern of dark and relatively normal peel is mindful of the stripes of a zebra.

In Florida the preharvest conditions commonly leading to postharvest zebra skin are trees in a drought stress followed by a sudden availability of soil moisture. The trees absorb water and the fruit become turgid. Distended epidermal cells in the ridges over fruit segments are most exposed and likely to be damaged by harvesting, handling and packingline machinery, especially brushes.

Only partial, usually not economic, control can be achieved by very careful handling. Delaying harvest for 5 to 7 days after the rain or irrigation that initiated zebra skin will usually allow the fruit turgor pressure to equalize for safe normal handling. Fruit sampling for normal handling during the susceptible time is advisable to judge the likelihood of the occurrence of this disorder. This disorder is increased by degreening, and is usually most severe on fruit with the greatest green color. Delaying harvest for better development of more orange color is another means of partial control of zebra skin.

Figure 93. Plugging of mandarins with a portion of the peel removed with the stem during harvesting.

Figure 94. Zebra skin with darkened areas over the segments of a tangerine.

FREEZING

Subfreezing temperatures can cause significant damage to citrus trees, leaves and fruit. Most citrus will tolerate temperatures as low as -2.2°C (28°F) with little or no damage if durations are not excessive. However, lower temperatures and longer durations will increase damage until the tree is eventually killed.

When temperatures are slightly below freezing, leaves may get a water-soaked appearance (Fig. 95). Such leaves usually recover. However, when leaves freeze until they are stiff and brittle, severe damage or death will likely follow. Following a freeze, quick leaf drop is a sign that twigs are probably not badly damaged. However, if leaves persist (Fig. 95), the twigs are usually badly damaged or dead.

Limb damage is more difficult to assess except when bark is split. Damage assessment should wait until new growth commences. Even then, appearances can be deceiving since there may be damaged areas below new sprouts. Pruning of damaged trees should not commence until the extent of damage is confirmed. Irrigation is of little value and fertilization should be reduced dramatically.

Citrus fruit injury can begin with as little as 2 to 4 hours below -2.2°C (28°F). The amount of damage sustained depends upon both the minimum temperature and the duration of freezing temperatures. Minor fruit injury will not be obvious externally and often is difficult to detect internally. Symptoms become more obvious as the injury increases. By slicing through an injured fruit, there will be a distortion of the segment membranes. Juice vesicles are often ruptured due to ice formation in the juice sacs. White hesperidin crystals may be deposited on membrane and vesicle surfaces. Later, the juice vesicles in injured areas will dry out and shrink (Fig. 95).

The extent of damage can usually be assessed shortly after a freeze by making successive cuts to see how far within the fruit freezing has occurred. Damaged fruit can, in most cases, be salvaged for juice, but must be harvested quickly since drying occurs rapidly and fruit drop will reduce the number of fruit harvested. Drops should not be used due to danger of microbial contamination.

Prevention of tree damage is by the use of warm sites and cold protection devices. Some fruit varieties may be harvested prior to cold weather. It is difficult to avoid tree damage in severe freezes.

Figure 95. Leaves remaining on trees (top) indicate severe freeze damage to twigs. Cold damage to grapefruit leaves showing a water-soaked appearance (center). Orlando tangelo fruit with distorted segment membranes and drying (bottom).

BLUE ALBEDO

Blue albedo is a condition where the albedo (white portion of the peel), and sometimes the vascular bundles, segment walls and occasionally juice sacs, have a distinct light blue color. The discoloration can vary from distinct vascular bundle patterns (Fig. 96) to an overall bluish color in the albedo. This condition has been observed in fruit from Arizona, Florida, California and Texas. As early as 1944 blue albedo was reported to be associated with stubborn disease. It has also been reported to be more prevalent when excessive fertilization or high salt content of shallow soils damage the roots.

In Florida this disorder has been seen in oranges, tangelos and especially grapefruit. Blue albedo is reported nearly every season, and is found frequently in some seasons. Consumers occasionally report this disorder as an alarming observation.

Blue albedo is nearly always reported in seasons with excessive rainfall, and the fruit is usually from groves with marginal water drainage. Trees under stress from excessive water may produce fruit with blue albedo. The blue color is a naturally occurring water soluble pigment, probably an anthocyanin similar to those found in blueberries and sweet cherries. The color persists in fresh fruit held at 4.4°C (40°F) and 21.1°C (70°F), but gradually fades in 3 weeks from chilled sections held at 4.4°C (40°F).

Control is effected through careful management. Avoiding excessive fertilization of shallow soil groves and selecting well drained groves for fresh fruit marketing in seasons with excessive rainfall will minimize the problem.

STYLAR-END RUSSETTING

This fruit blemish which is not very common most years, is characterized by a stippling or russetting of the stylar ends of the fruit (Fig. 97). When examined with magnification, the blemish will appear to be a linked series of very slightly raised corky lines in a netted pattern. Although it may appear on many varieties, stylar-end russetting is most often reported on navel oranges. The russetting is often confused with rust mite injury unless scrutinized closely. The injury was believed to have been caused by feeding of the citrus rust mite, but this assumption was in error. Curiously, the cause of the malady still remains unknown.

Figure 96. Blue albedo vascular bundles in grapefruit peel segment.

Figure 97. Stylar-end russetting on a navel orange with characteristic netted pattern.

FRUIT CHIMERAS

There are 2 types of chimeras, periclinal and sectorial. The sectorial chimera (Fig. 98) is a fairly common occurrence. This is a slight genetic mutation of the tissue, and as the name implies, corresponds to 1 or more sectors of the fruit. A chimera may be a darker or lighter colored stripe, or a raised or sunken area on the peel. Chimeras are cosmetic and should not be of great concern. Chimeras sometimes have resistance to certain insects or diseases, or have better color or quality. Plants regenerated by tissue culture from these segments would likely have these desirable characteristics.

The periclinal chimera affects 1 or more layers of the fruit. The peel or the pulp is most likely affected in entirety and unlike the sectorial chimera, there is no obvious contrast and differences are difficult to detect. The Thompson pink grapefruit is a periclinal chimera of Marsh and seed will always produce white fruit since only the flesh is chimeric.

Since chimeras are slightly different genetically, trees should not be propagated from plants which produce large numbers of chimeric fruit. Chimeras are also grade-lowering factors and, as such, are not desirable on fruit destined for the fresh market.

HAIL INJURY

Damage from hail is fairly common in Florida but rarely affects a very large growing area. The amount of damage sustained is a function of the duration and intensity of the storm, the size of hailstones and the stage of development of trees and fruit. Injury from hail results from physical damage of the hailstones pummeling fruit, leaves and twigs. Twigs will have bark which is bruised and split. Leaves will be shredded and torn (Fig. 99). Fruit will be pitted and damaged areas may later become corky as the fruit heals (Fig. 99). Severely injured fruit may be knocked off the tree or fall at a later date due to injuries sustained. Peel damage may also result in later injury due to invading decay organisms.

The appearance of the damage on the tree is practically unmistakable in the field. Diagnosis may be more difficult on harvested fruit as similar appearing injuries can occur from twig damage, rough handling or bird pecks.

Figure 98. Sectorial chimeras on navel oranges.

Figure 99. Hail injury results in shredded, torn leaves and corky pitted spots on fruit.

CREASING

Creasing (Fig. 100) results when a separation or crack develops in the albedo (white spongy area of the rind) causing a furrow or crease to occur in the overlying flavedo (colored portion of the rind). These furrows are random in occurrence, with no particular orientation. Most sweet oranges can be affected as can Temple and other types of citrus fruit. The problem is sporadic and usually affects only a fraction of the fruit produced on a tree.

Creasing may be related to temperatures which occur during the period of maximum peel growth and development, but there are insufficient data to corroborate this as the cause. The problem is most severe on potassium deficient fruit that are often thin skinned. Maintenance of adequate potassium levels is therefore important to help minimize creasing. Phosphorus levels seem to have little effect on creasing incidence, but high levels of nitrogen increase creasing. Adherence to a program of adequate potassium and avoidance of excess nitrogen levels should help to minimize the problem.

Where creasing is a problem, it seems to be accentuated as the fruit mature. Therefore, early harvest will help to minimize the problem.

FRUIT SPLITTING

Fruit splitting is primarily a problem in the fall under Florida conditions (Fig. 101). The exact cause is unknown but is believed to be a problem of water relations and peel thickness. Damage often occurs when trees take up water from rain or irrigation and the fruit expands, bursting the peel in a crack across the bottom. Less often, the fruit may split in other places longitudinally and occasionally even equatorially.

Other contributing factors may include potassium and copper deficiency, both of which result in reduced peel thickness. Fruit on young trees seem to have a greater propensity to splitting than do fruit on mature trees. This is probably associated with water uptake and the increased water buffering capacity of the larger trees. While the problem cannot always be avoided, maintenance of optimum growing conditions and furnishing adequate, but not excess, levels of nutrients should help to minimize split fruit.

Figure 100. Creasing of Dancy tangerine showing furrows on the surface (left) and with missing tissue in the albedo (right), where flavedo has been removed for clarity.

Figure 101. Longitudinal fruit splitting of Hamlin oranges.

GRANULATION OF FRUIT

Fruit which is dry or ricey is said to be granulated (Fig. 102). The terms are often used interchangeably and refer to the fact that the fruit has dried out. Freezing injury (page 186) and sunburn (page 198) also cause citrus fruit drying. Other major causes include over maturity, a lack of water, excessive tree vigor, severe mite damage and cool, dry, windy conditions. Since water will be drawn from the fruit to supply the needs of the tree, keeping trees well irrigated will help keep fruit from drying out. Likewise, prevention of high populations of mites will keep peels from being damaged with subsequent loss of water through the damaged skin. Weather conditions cannot be controlled, but conditions which promote rapid water loss can be countered with irrigation.

Perhaps the most frequent cause of fruit granulation is excess tree vigor which occurs most often with young trees. Such trees are grown rapidly to increase tree size and this compromises fruit quality. Best fruit quality and absence from granulation will occur after young trees slow down their growth rate, usually at an age of 4 to 5 years.

LIGHTNING INJURY

Lightning injury is fairly common in Florida during the frequent thunderstorms. In some cases, only minor injury occurs to a single tree but there are instances where many trees have been killed by a single bolt. On individual trees, damage may range from slight twig injury and subsequent defoliation to death of the tree. Most often 1 or more branches are killed and there is often a lightning path of damaged or dead bark extending from the top of the tree down to the ground. When damage is severe, the limbs or even the entire tree will die due to disruption of the vascular system.

Damage to twigs from lightning provide an unmistakable diagnostic tool. A small area around each leaf petiole, thorn or node will remain green while most of the rest of the bark on the twig will be killed (Fig. 103). These islands or green areas are unique to lightning injury and occur on most trees which have been hit. Where damage is substantial, the tree should be removed and replaced. If damage is less severe, damaged limbs could be pruned out and the tree should recover.

Figure 102. Granulation of a Valencia orange fruit with drying of the juice vesicles.

Figure 103. Lightning injury with green islands along a twig (Courtesy R. L. Phillips).

WIND SCARRING

Fruit are often injured by abrasive contact, most often leaves, twigs or thorns. While injury can occur at any time, most damage probably comes during the windy months of spring while fruit are rather small. Injured areas grow with the fruit, making minor blemishes appear quite large at fruit maturity. Most of the injuries suberize quickly and damage is primarily cosmetic. Scars caused by injuries are grade-lowering factors which can reduce the value of fruit for the fresh market.

When small, immature fruit are injured, the damaged surfaces develop a rather smooth, thin, tannish area as the fruit continues to enlarge (Fig. 104). Injuries which occur near fruit maturity are less frequent and less obvious as the fruit has thicker peel and the damaged area will not enlarge greatly. Late injuries are most often due to thorn pricks or limb damage and may be covered with a dark, corky scar tissue. Sometimes injury sites on fruit at any stage of development are invaded by microorganisms, causing further damage or fruit drop.

Other scarring injuries may occur from damage due to various pests. Grasshoppers, katydids and crickets will feed on the peel, causing a depressed area below the surface level of the adjacent, uninjured peel. Several species of thrips may cause feeding injury. One species produces feeding injury around the distal end of the fruit, often in the shape of a ring. Similarly, another species may cause injury around the area where 2 fruit are in contact with each other. Yet other species of thrips may be serious problems in other citrus-growing regions, especially in arid areas. Populations of thrips in Florida have not been damaging enough to warrant control, although recent evidence indicates that thrips may be more of a problem than previously believed.

Wind scar injury is controlled by avoidance, which is difficult. Regular pruning to keep dead wood to a minimum and the use of peripheral windbreaks will help to minimize such injuries.

Figure 104. Irregular smooth tan pattern of wind scar on the surface of an orange.

SUNBURN

The intense radiation of the Florida sun sometimes damages leaves, stems and fruit. Modest damage to leaves sometimes occurs when leaves are exposed to the sun after having been previously shaded (Fig. 105) and may result in gum spots.

Gum spots on the leaves is most often due a change in the leaf position, exposing it to strong sunlight. The damage consists of irregularly-shaped dark brown areas with a smoother, slightly raised surface. A hard, gum-like material covers the surface layer of cells. Often, the area around the gum spots is somewhat lighter in color than areas more removed from the spots. The problem at first glance is somewhat like melanose (page 146) or greasy spot (page 142), but close inspection will easily distinguish the difference.

Fruit damage may occur in fruit which is exposed to long durations of sunlight, especially on the shoulders of the tree in the southwest quadrant. When such damage occurs it is most often on thin-skinned varieties, especially those which bear fruit which extends beyond the leaf canopy. The Honey tangerine (Murcott) is 1 of the most susceptible due to its growth habit. Damage is manifested by drying of exposed portions of the peel and the pulp and juice beneath it. Damaged areas also will usually have an advanced color break (Fig. 105). Usually only the most exposed portions of the fruit are affected which distinguishes this problem from other types of granulation (page 194).

Damage to twigs and limbs most often is the consequence of pruning which opens up shaded portions of the tree to sunlight. Just as people sunburn, the previously shaded portion of the tree may be burned by the sun. Bark can be injured or killed and may later slough off. Avoid pruning during the summer months to help prevent this injury. If summer pruning is necessary, protect vulnerable limbs with white latex paint or whitewash to reflect the sun's rays and keep temperatures of the limbs below damaging levels.

Figure 105. Sunburn of leaves with gum spots (left) and severely damaged leaf (center). Honey tangerine fruit (Courtesy D. Tucker) exhibit advanced color in sunburned areas.

HERBICIDE INJURY

Damage to citrus trees may sometimes occur as a result of improper application of herbicidal materials. As a consequence, some herbicides may be taken up by the citrus tree roots, leaves and green wood. When damage occurs from root uptake, the problem is usually due to misapplication resulting in excessive rates of applied chemicals.

Poor spray distribution leading to excessive overlap can result in localized higher rates of application with resulting possible tree damage. Inadequate agitation and distribution of the herbicide in the tank can lead to uneven application rates, with possible tree damage since spray concentration reaching the ground may vary. Wettable powder formulations should be premixed with water as a slurry prior to placing in tank so that agitation equipment can achieve satisfactory mixing. Flowable, true liquid and emulsifiable concentrate formulations should also be agitated thoroughly.

In some cases, the damage results from actual application of herbicides to the citrus tree instead of the target weeds. This may be due to improper nozzle placement or equipment setup or could be the result of drift. Spray drift can occur if application pressures are too high, booms are improperly shielded or if applications are made under windy conditions. Repeated spraying of tree foliage and green wood with contact and some systemic herbicides will lead to increasing levels of tree damage. The addition of certain surfactants to some residual herbicides will increase the contact activity and leaf injury symptoms.

Symptoms of growth regulator herbicide injury are usually manifested as leaf distortion such as curling or strapping (Fig. 106) and growth anomalies such as vigorous growth or a reduction in internode length. Contact herbicides cause desiccation injury by burning tissue in spots (Fig. 106), or in severe cases, even larger surface areas. Such damage is most often found on green tissue near the ground where errant spray applications may have impinged on tree surfaces. Residual herbicides usually cause damage through root uptake. Symptoms vary with chemicals but are most often some sort of foliage chlorosis that is associated with leaf veins (Fig. 106). Some chemicals can cause irregular, bleached patterns. In some cases, leaf distortion and dieback may occur.

Figure 106. Herbicide injury on leaves showing curling or strapping (top), contact burning (center, Courtesy D. Tucker) and (bottom) chlorosis of veins (Courtesy D. Tucker).

SPRAY AND DUST INJURIES

Citrus foliage and fruit are sometimes injured by sprays and dusts. There can be many reasons for and many symptoms of such injuries resulting from excessive rates, equipment problems, unfavorable climatic conditions and most often, mixing several chemicals of unknown compatibility. Most spray and dust injuries can be characterized by discoloration and/or burning of the tissue and in severe cases, abscission of the fruit and/or leaves and even twig dieback. Burns are most likely to appear on exposed surfaces where concentrate sprays impact tissue, but may occur at the bottom portions where dilute spray runoff accumulates and concentrates. Much of this sort of damage can be avoided by critical evaluation of the spray program. Careful record keeping can help unravel problems associated with later spray damage. Several common spray/dust injuries are discussed below.

Oil blotch injury is usually manifested as nearly circular (often coalescing) light brown areas concentrated on the outside of injured fruit. Oranges, tangerines and tangerine hybrids are most susceptible to this injury, especially when the fruit is between 3/4 and 1-1/2 inches in diameter (Fig. 107). Oil blotch injury is superficial and usually consists of a smooth single layer of corky cells. Judicious use of highly refined spray oils and spraying under proper conditions will help to minimize oil blotch.

Copper-oil Spray Injury results sometimes when oil is used with copper or other metallic elements. This peel injury is characterized by circular spots corresponding to the pattern of drying spray droplets. The spots are likely light brown or grayish, have a corky appearance and slightly sunken (Fig. 107). The use of high quality spray materials will minimize this problem.

Sulfur Burn occasionally results in fruit damage on the exposed side (Fig. 107). The injured area is irregular in shape and size but most often is 1 to 2 inches in diameter. The affected area is dark brown and may be slightly lighter in the center. There is not a distinct margin to the damaged area and there may be smaller, darkened spots surrounding the main area of damage. There is always a risk of sulfur burn when the spray or dust is applied during periods of hot, dry weather and when used at high rates.

Figure 107. Spray and dust injuries may be concentrated (top) at bottom of fruit (Courtesy D. Tucker), streaked (center), or a large discolored area (bottom) as with this sulfur burn (Courtesy R. L. Phillips).

SALT OR FERTILIZER BURN

Damage to citrus from high levels of salt is manifested in several ways. Most often the problem is noticed in the leaves as irregular dead brown areas, usually near the margins and tips (Fig. 108). The damaged areas are frequently invaded by a secondary fungus, *Colletotrichum gloeosporioides*, (pages 142, 150) which could lead the casual observer to believe that the fungus caused the problem. However, this strain of *Colletotrichum* is an opportunistic organism that is basically a symptomless invader of damaged tissue. Another strain of this fungus causes the disease known as postbloom fruit drop (page 150). Acute salt damage will result in defoliation of all or part of the leaves and in the worst cases, dieback and even death of the tree. If a chronic salt condition exists, affected trees will be sparsely foliated and usually show only mild symptoms. Root damage will nearly always accompany salt injury but this damage is not obvious to the observer unless roots are excavated or the tree is removed and examined.

Salt injury is usually caused by exposure due to salt spray, salt water influx in coastal areas, or irrigation of trees with water containing high levels of dissolved salts. Excessive application of fertilizer may also result in this type of damage. Combinations of any 2 or all of the above are possible but the effect is the same.

Control of the problem is through avoidance. Leaves which are damaged will not recover and will likely fall prematurely. Injured roots may regenerate, but only if conditions improve. Sites which are prone to salt deposition from ocean winds or which are subject to tidal influx should be avoided. Irrigation water should be checked regularly to insure salt levels do not become excessive. Where salt levels exceed 1000 ppm, irrigation should be undertree only and ideally, applied only during periods of low evaporation. Fertilization should be monitored to insure that toxic excesses are avoided. The use of low salt index fertilizer materials should be considered and fertilizer applied as frequent, light applications as opposed to only a few larger applications.

Figure 108. Salt or fertilizer burn with resultant irregular dead brown areas along the margins and tips of leaves (Courtesy D. Tucker).

NITROGEN DEFICIENCY

Nitrogen deficiency symptoms occur whenever the nitrogen supply in the soil is inadequate to support normal growth and fruit production. Nitrogen deficiency commonly would be found in almost all citrus growing regions of the world if it were not for the almost universal practice of supplying nitrogenous fertilizer.

Nitrogen deficiency ordinarily can be distinguished from many other deficiencies by the appearance of the foliage, yellowing over the entire tree, and the absence of any distinct leaf patterns. As trees become deficient in nitrogen, leaves become lighter green in color. If the deficiency is not corrected, the leaves generally become a distinct yellow-green with some leaves lacking any green coloration (Fig. 109). In advanced cases, canopy foliage becomes sparse and twig death may occur, and as leaves age or senesce they become yellow or bronze-yellow (Fig. 109). Yellow vein chlorosis (page 218) accompanying girdling or root injury has been ascribed to and is associated with nitrogen deficiency.

Nitrogen deficient trees have reduced fruit yields. These fruit may be bland in flavor due to low sugar and acid. Fruit also may have poorer external and internal color. Fruit from nitrogen deficient trees also have been found to be more prone to decay in storage.

Nitrogen deficiency symptoms may be expressed at any time. When nitrogen deficiency symptoms occur in the early spring, it is probably due to low soil temperatures and root inactivity with resulting reduced nitrogen uptake. Late summer and fall symptoms can be the result of depletion of the soil nitrogen reserves due to excessive summer rainfall. Varieties such as Nova, Minneola and Orlando tangelos are particularly sensitive, and can become nitrogen deficient in the fall and winter.

Nitrogen deficiency readily can be corrected by the application of nitrogen fertilizers. Readily soluble forms applied to the soil and watered in can provide a visible response in as little as 1 to 2 weeks. Nitrogen deficiency also can be corrected by foliage application of low biuret urea.

Excessively high nitrogen fertilization ultimately can reduce yields, especially in groves with no supplemental irrigation. Excessive nitrogen has been found to result in fruit that are smaller, more coarse textured, and greener with high sugar and acid levels.

Figure 109. Stages of nitrogen deficiency (top) with a healthy leaf on left to a severely affected leaf on right. Bronze-yellow color of advanced nitrogen deficient leaves (bottom).

PHOSPHORUS DEFICIENCY

Phosphorus deficiency symptoms rarely are seen in commercial citrus groves in Florida where phosphate fertilizer has been applied. Phosphorus is used by the tree in smaller quantities than nitrogen and potassium and is retained against leaching loss more readily in the soil profile. Plantings on virgin flatwoods soils likely will be phosphorus deficient if phosphorus fertilizers are not applied.

In the field, phosphorus deficient trees frequently have leaves which are a dull bronze in color. Fruit symptoms are more obvious, being rough in external texture with a thick rind and sometimes a hollow core (Fig. 110). Fruit from deficient trees are high in acidity. Phosphorus deficiency has been found to be associated with stunted growth of citrus seedlings in the field nursery and greenhouse, particularly where mycorrhizal fungi have been eliminated by fumigation. Soil aeration following heavy phosphate applications to phosphate starved seedlings has been shown to partially or totally restore normal growth.

POTASSIUM DEFICIENCY

Potassium deficiency results from an insufficient supply of potassium in soils, or because of restricted root uptake as a result of antagonistic effects of excess calcium and magnesium in calcareous soils. The most characteristic deficiency symptom is the production of small, smooth fruit (Fig. 111) with thin peel and low acid content, which usually mature earlier. Such fruit usually have more desirable color than those on trees receiving adequate or excessive amounts of potassium. Fruit adequately supplied with potassium tend to have coarser external texture. With severe potassium deficiency, there is twig dieback and a reduction in leaf growth. Leaf deficiency symptoms rarely are found in Florida. Where deficiencies have been induced, leaves appear bronze in color and distorted in shape. Premature fruit drop, splitting (page 192), creasing (page 192), plugging (page 184) and yield loss are other potassium deficiency symptoms.

Potassium deficiency usually is corrected by applying potassium fertilizers to the soil, and may be corrected by applying potassium nitrate sprays to the foliage. Foliar application of potassium nitrate may be used where excess soil calcium and/or magnesium is present.

Figure 110. Marsh grapefruit with phosphorus deficient symptoms (left) with tick rind and hollow core and a rough external texture (Courtesy R. L. Phillips).

Figure 111. Pineapple orange with potassium deficiency symptoms (bottom) of small smooth textured fruit (Courtesy R. L. Phillips).

MAGNESIUM DEFICIENCY

Magnesium deficiency of citrus is encountered primarily in acidic sandy soils from which magnesium may be readily leached. Magnesium deficiency also occurs in calcareous soils which are low in magnesium or where excess calcium interferes with magnesium uptake.

Young leaves on trees growing on soils low in magnesium may show no magnesium deficiency symptoms in the spring, but magnesium deficiency patterns may appear in late summer or fall on the mature leaves. An irregular yellowish-green blotched area starts at the leaf margin and spreads interveinally toward the midrib. Such areas enlarge until only the tips and bases of the leaves are green as inverted V-shaped areas on the midrib (Fig. 112). Under unfavorable environmental or cultural conditions, leaves with acute magnesium deficiency symptoms become completely yellow and eventually drop. Defoliation can be heavy on magnesium deficient trees, especially on limbs carrying a large crop. Limbs on the same tree bearing little or no fruit often exhibit no symptoms. Fruit yield eventually will be reduced. Seedy orange and seedy grapefruit varieties usually are more susceptible to chronic magnesium deficiency than other citrus cultivars.

If leaf magnesium deficiency symptoms occur, magnesium sulfate or oxide should be applied in the fertilizer, and the rate of magnesium should be increased up to 30% of the nitrogen rate until symptoms are no longer present in mature leaves or subsequent flushes. Citrus growing on soils high in calcium often requires above normal levels of magnesium fertilizer for satisfactory tree nutrition. In cases where soil-applied magnesium fertilizer is ineffective, leaf magnesium concentrations may be increased through foliar applications of magnesium nitrate.

Figure 112. Grapefruit leaves with mild (left) to severe magnesium deficiency with the characteristic V-shaped green areas along the midrib.

ZINC DEFICIENCY

Zinc deficiency is among the most commonly observed citrus deficiencies. Low zinc content in leached sandy soils, soil alkalinity in high pH calcareous soils, and alkaline-forming nitrogen fertilizer sources may induce symptoms. Differences occur among rootstocks, with Carrizo and other trifoliate hybrids being more susceptible, especially in high pH soils. Zinc deficiency may occur in all citrus varieties, being most severe on oranges and least on tangerines. Pineapple and Valencia oranges usually are more susceptible than early season oranges.

Deficiency symptoms develop gradually, appearing first on a few young leaves and more on the sunlight side of the tree. Zinc deficiency is an early diagnostic symptom of blight (page 124). In early stages the leaves are of normal size and shape with only small blotches of yellowing between veins, similar to iron or manganese deficiencies and 2 or 3 of these deficiencies may occur together. Zinc deficient leaves occasionally exhibit small green dots in the yellowed areas, not seen with manganese deficiency. Narrow, irregular green bands extend along the veins with the most green color along the midrib (Fig. 113). In contrast the interveinal areas are mainly pale yellow, ranging in some cases to a cream color. Severely deficient leaves may become entirely yellow with irregular green areas between the veins and interveinal areas. Leaves may be small and narrow with pointed tips (little leaf or mottle leaf) (Fig. 113). Decreased stem elongation and shortened internodes result in small leaf clusters and rosetted twigs. Advanced deficiency results in weakened terminal growth, leaf drop, and twig death, and eventually in yield reduction. Fruit are usually small, sometimes misshapen, and light yellow, and the pulp tends to be dry and insipid.

As transient zinc symptoms are not associated with yield reductions, control measures are only suggested in Florida when symptoms persist. Many groves which have not received zinc for many years show no symptoms. While both foliar and soil applications of zinc have corrected deficiencies, precipitated zinc sulfate or oxide sprays applied postbloom to trees with fully expanded spring growth flushes are most effective. Latter treatments are not as effective. Liquid chelate formulations should not be tank mixed with spray oils for postbloom applications as injury to fruit may result.

Figure 113. Zinc deficiency from mild (top left) to severe narrow, pointed tip leaves (top right and bottom, Courtesy I. Stewart).

MANGANESE DEFICIENCY

Manganese deficiency occurs in both acid and calcareous soils, probably because of leaching losses in the former and very low solubility and consequent unavailability in the latter. Manganese deficiency symptoms commonly found in Florida and other citrus growing regions of the world are characterized by dark green bands along the midrib and main veins of leaves surrounded by light green interveinal areas. In mild cases, symptoms appear on the new flush, frequently becoming less pronounced or disappearing during the life of the leaf as the light green areas become darker. As the deficiency increases the light green color of the interveinal areas may give way to a bronze appearance which will persist (Fig. 114). Leaf size generally is not affected, and there appears to be little yield reduction until the deficiency becomes acute enough for defoliation and twig dieback, a rare occurrence in Florida.

Unless the symptoms are typical, and the foliage free of zinc and iron deficiency patterns, manganese chlorosis may be difficult to identify. In such cases, the cause of the problem can best be identified through chemical leaf analysis. No fruit symptoms have been reported.

Soil applications of manganese, preferably in the sulfate form, are effective when trees are growing under acid but not under calcareous conditions where foliar sprays must be used. Applications of manganese are not considered necessary or economically feasible unless manganese deficiency symptoms are present.

Figure 114. Manganese deficiency with dark green bands along the midrib and bronzing (right) in more severely affected leaves.

IRON DEFICIENCY

Soil interactions are important when evaluating iron deficiency. Although deep sandy soils of central Florida are relatively low in iron, deficiency symptoms rarely are observed in commercial groves. Deficiency symptoms in white sandy soil production areas probably are due to a lack of iron. Along Florida's east coast where some soils naturally high in iron also contain high amounts of calcium, with pH values between 7.0 and 8.5, iron deficiencies usually are due to unavailability. In other cases deficiency in plants has been attributed to antagonism of other heavy metals, primarily copper, accumulated in acid soils. High copper levels directly interfere with iron uptake. Deficiency symptoms also may occur following root damage from inadequate drainage or nematode infestations. Foliar symptoms appear first on new leaves and in mild cases iron deficiency easily can be recognized by leaf veins that are darker green than interveinal areas (Fig. 115). These chlorotic symptoms may become less evident as leaves mature. In more severe cases, interveinal areas become yellow and entire leaf blades eventually become ivory in color (Fig. 115). Severely deficient trees may be partially defoliated with eventual twig and canopy dieback. Fruit from deficient trees are low in soluble solids and high in acid, with oranges developing a lemon yellow color.

Iron deficiency may occur in combination with zinc and manganese deficiencies since deficiencies of these elements are accentuated in high pH soils (pH greater than 6.0). Visual recognition of the specific deficiency may be difficult. Leaf vein green color bands are wider for zinc or manganese deficiency than for iron.

Iron deficiency can be corrected most satisfactorily by soil applications of iron chelates, selected for soil pH. In calcareous soils, repeated applications of chelates may be needed to prevent recurrence. Foliar iron sprays are usually ineffective for moderate to severe deficiencies. Foliar applied iron sprays partially may regreen foliage with minor symptoms, but new flushes following application may express symptoms. Iron chelate (EDTA) foliar sprays should not be used since they can cause foliage and fruit burn. Acidification of calcareous soils with acid forming fertilizers or sulfur eventually may alleviate the condition by making iron available, but this is a long term and possibly impractical procedure.

Figure 115. Progressive iron deficiency (top) with characteristic dark leaf veins and yellow interveinal areas, and (bottom) eventually an ivory color (Courtesy D. J. Banks).

BORON DEFICIENCY

Some Florida soils are generally low in boron, and deficiency symptoms on citrus would be widespread without regular boron applications. Boron deficiency is most often manifested in fruit that are usually small, lumpy, misshapen, and hard with a thickened, occasionally cracked rind. Gumming on the outside of the fruit and brown areas in the albedo (Fig. 116), central core, and fleshy parts of the fruit may also be present. Seeds may be shrivelled and have darkened or entirely missing seed coats. Affected fruit can be dry and low in sugar content. Fruit drop can be excessive. Trees with symptoms of boron deficiency usually lack vigor. Leaf curling, vein corking and enlargement (Fig. 116), and splitting and gumming of twigs and branches are symptoms which may occur on the tree.

Boron may be applied as borax or other soluble borates to citrus groves as needed in either mixed fertilizer or in spray. However, both ground and spray applications of this element should not be applied in the same year. Apply boron as part of the fertilizer maintenance program at an annual rate equivalent to 1/300 (one three-hundredth) of the nitrogen rate. When applied as a foliar maintenance spray, 1.25 pounds of soluble borate, containing 20% boron per 500 gallons may be used. It is suggested that the amounts be doubled when deficiency symptoms are present.

YELLOW VEIN CHLOROSIS

This type of leaf pattern frequently may be attributed to the girdling of individual branches or tree trunks due to a number of causal factors including water damage, foot rot, chemical injury, lightning, gopher, armadillo, termite or ant damage, or physical damage due to mechanical cultivation equipment. Yellow vein chlorosis also may occur on twigs and branches with the onset of cooler weather in the fall and winter probably due to lack of nitrogen uptake from the soil, a condition known as winter chlorosis. Certain scion varieties such as Orlando tangelo and rootstocks, particularly Swingle citrumelo, are most often associated with this condition.

In leaves showing yellow vein chlorosis, the midribs and lateral veins and a band of leaf tissue bordering them become yellow while the rest of the leaf remains a normal green color (Fig. 117). In contrast, the entire leaf turns yellow when nitrogen is too low.

Figure 116. Boron deficiency of citrus fruit may have dark areas in the albedo (white area of the peel) and corresponding raised area (not shown) of the peel (left). Leaves of boron deficient citrus (right) exhibit leaf curling and vein enlargement.

Figure 117. Yellow vein chlorosis with yellow veins and midrib progressing a completely yellow leaf (Courtesy R. L. Phillps) and whole tree (Courtesy R. L. Phillips).

COPPER DEFICIENCY

Copper deficiency is caused by lack of available copper in the soil. Excessive nitrogen fertilization also has been considered to induce copper deficiency. While once a common occurrence in Florida, copper deficiency is now rare, especially in the older citrus growing areas where ample levels of copper have accumulated from soil applications and sprays. However, on the shallow, poorly drained, previously unfertilized soils of the flatwoods areas, copper deficiency often develops if the element is not regularly included in the fertilizer or nutritional sprays.

Early evidence of copper deficiency is the occurrence of unusually large, dark green, leaves on long soft, angular shoots. These leaves commonly are irregular in contour, usually with a bowing up of the midrib. Affected trees appear unusually vigorous and soft shoots sag at the tips or become S-shaped (Fig. 118). Before young affected twigs attain their maximum length, blister-like pockets of clear gum frequently develop between the wood and bark, usually near the leaf nodes. As these twigs mature, reddish-brown eruptive lesions occur in the outer portion of the bark (Fig. 118). Considerable portions of the bark surface may be stained brown by gum deposition, and leaves also may develop stained areas. Multiple buds may form at the nodes. Severely affected twigs commonly die back from the tip and new twig growth occurs from the multiple buds, eventually resulting in a bushy cluster of dead twigs.

Fruit symptoms include dark reddish-brown, gum-soaked irregular shaped eruptive lesions on the peel surface (Fig. 118). Lesions vary in size from a few small specks to numerous large areas, sometimes coalescing to cover most of the fruit surface. Symptoms may occur from when fruit are 1 month old until they nearly attain maturity. In mild cases, the lesions may be mistaken for melanose (page 146). Copper deficiency and melanose symptoms may be distinguished usually by the greater depth and sometimes spongy consistency of the copper deficiency lesions compared to the hard, superficial nature of the melanose lesions. Other fruit symptoms of copper deficiency of citrus include gum in the rind and around seeds and fruit cracking. Fruit from copper deficient trees tend to be low in soluble solids, acid, and juice.

Figure 118. Copper deficient trees appear vigorous with S-shaped young shoots (top), and twigs (center) develop gum between the wood and bark with lesions near the leaf nodes (Courtesy D. Tucker). Fruit of copper deficient trees exhibit dark lesions on the surface (bottom).

MOLYBDENUM DEFICIENCY

Molybdenum deficiency of citrus used to be referred to as yellow spot in Florida. This seldom-found deficiency is caused by insufficient uptake of molybdenum from sandy soils. Symptoms appear on leaves as large yellow interveinal chlorotic spots (Fig. 119). Although the spots may be found at any time during the year on deficient trees, they usually appear first as water-soaked areas on old leaves in early summer. Following the summer growth flush, bright yellow spots develop on older leaves with deposits of brown gum on the lower leaf surfaces. With severe deficiency, the spots will increase in size until they coalesce, and during the winter many of the affected leaves may drop. Symptoms seldom are found on the fruit except when deficiency is severe. Fruit grown under severe deficiency conditions may have large irregular brown spots with a yellow halo with the discoloration only in the outer peel and not penetrating into the albedo. The fruit spots, like the leaf spots, are found more commonly on the parts of the trees most exposed to the sun. In older groves the symptoms may be found over the whole trees. Deficiency symptoms are more severe on trees on grapefruit rootstock than those on other rootstocks.

Although the deficiency can be corrected with sodium molybdate sprays, the neutralizing of highly acid soils with limestone may be a more effective method of correcting this deficiency. If molybdate sprays are the treatment of choice, applied in the summer shortly after the symptoms occur or up until October, the leaves will regreen and the spots disappear from the upper surface. Most of the gum on the lower surfaces of the leaves will be replaced by black spots consisting primarily of cork cells. One spray generally is sufficient for three or more years.

Figure 119. Molybdenum deficient leaves develop large interveinal yellow spots (Courtesy I. Stewart).

BIURET TOXICITY

Biuret toxicity symptoms on leaves can develop following the application of biuret-containing nitrogen fertilizers to citrus trees. Biuret, an impurity of some urea fertilizers, is formed when urea is melted in the pelletizing procedure. Trees receiving foliar sprays or soil applications with biuret-containing fertilizer can develop the symptoms. Foliar applications of urea containing biuret of 0.25% or greater will cause symptoms of biuret toxicity while soil applications of urea containing 2.5% biuret or greater will cause symptoms of biuret toxicity.

Leaf symptoms appear as irregular, yellowish-green interveinal chlorotic areas (Fig. 120), appearing first at leaf tips and, depending on the amount of biuret applied, spreading over the entire outer areas of the leaf surface. In severe cases, only the midribs and parts of major veins remain green. Leaves with necrotic areas may appear in the most severe cases and these leaves eventually may drop. Young leaves and trees usually are more susceptible than older leaves and trees. Symptoms of chlorosis can develop within 3 weeks of application in the summer, while in winter a longer period may elapse before symptoms appear.

Symptoms of biuret toxicity may be confused with those of boron, perchlorate, and some herbicides. However, the light yellow chlorotic areas and lack of gumming particularly are characteristic.

While symptom development may be avoided by using low biuret urea sources, no treatment is known to alleviate leaf symptoms once developed. Leaf flushes developing after the appearance of affected foliage usually are normal, indicating a reduction of the level of biuret in the tree and/or soil.

Figure 120. Biuret toxicity of citrus leaves is characterized by interveinal yellowish-green areas starting at the tip (Courtesy D. Tucker).

BORON TOXICITY

Boron toxicity of citrus in Florida usually results from excess boron in spray or dry fertilizer applications. An early stage of boron toxicity in the citrus leaves is a yellowing at the tips. These chlorotic areas often extend down the edges of the leaves and the yellow and green portions blend creating a mottled effect (Fig. 121). The leaf margins and tip may develop dead, scorched appearing areas and leaf drop occurs. Gumming excretions on the lower surface of the leaves (Fig. 121), also described as brown resinous spots, are conspicuous, distinguishing features of boron toxicity. If toxic levels of boron are still present after partial defoliation, new leaf flush also will show symptoms. With severe cases, the leaves of successive flushes are usually small and almost white colored and these twigs often die. Symptoms usually become less pronounced with each successive flush; however, in extreme cases total defoliation and tree death can result. Because boron accumulates progressively as citrus leaves age, the apical mottling is usually more pronounced in late summer and fall on the oldest leaves.

The susceptibility of citrus to boron toxicity varies with rootstock and scion variety. Grapefruit, for example, has been shown to absorb less boron when grown on sour orange than on trifoliate rootstocks. Of the scion varieties, lemon is more susceptible than sweet orange, with other varieties being intermediate. Toxicity due to high levels of boron in the soil can be decreased by frequent flushing of the root zone with irrigation water of low boron levels. Heavy applications of alkaline fertilizers and liming of acid soils also have been found to be effective. Boron toxicity can be prevented by exercising care in applying the recommended amount of the element as the range between sufficient and excess is small.

Figure 121. Boron toxicity of grapefruit leaves with a mottled yellow effect and gumming on the lower surface (right, Courtesy R. C. Koo).

FLUORIDE TOXICITY

Fluoride toxicity in citrus trees is caused by exposure to gaseous airborne fluorides. This condition was reported in Florida groves in the early 1950's in the vicinity of phosphate manufacturing facilities from which gaseous fluorides were emitted. Symptoms generally are more severe in groves in the path of prevailing winds from such facilities, and especially in groves relatively near them.

A characteristic chlorotic leaf pattern is associated with fluoride toxicity in citrus. Such chlorosis is most intense at the tip and margins of the leaves and extends, with diminishing intensity, into the interveinal areas (Fig. 122). In more severe cases, tip and marginal leaf burn occurs. A reduction in leaf size is often noticeable in groves exposed to relatively high concentrations of airborne fluorides for an extended time. The chlorotic pattern in severe cases may resemble that of boron toxicity without the gumming. Grapefruit are more susceptible to fluorides than oranges.

While an economically feasible solution to fluoride toxicity in existent groves is not known, new plantings should be located in areas where airborne fluoride levels do not exceed normal levels.

PERCHLORATE TOXICITY

Yellow tipping of citrus leaves is caused by perchlorate impurities in certain nitrate of soda or potash fertilizer sources. It does not occur when synthetic sources are used. The first symptom of this toxicity is a bright yellow-orange color on the tips of older leaves (Fig. 123). As the condition becomes more severe, the orange color may spread over approximately half of the leaf surface. The veinal areas usually remain green, only in advanced cases becoming chlorotic. The chlorosis appears first on leaves with the greatest sunlight exposure, most commonly in tree tops in canopied groves, and throughout the canopies in open groves. In some cases, partial defoliation may result.

Yellow tipping may be confused with boron toxicity and biuret injury. Chlorosis from boron excesses are usually dull yellow in color, and gum spots are present on the lower leaf surface. Light yellow chlorotic leaf areas are representative of biuret toxicity.

Grapefruit are more likely to develop yellow tipping than other citrus. Affected leaves remain chlorotic until they drop. No means of correction is known after the condition has developed.

Figure 122. Chlorotic leaf pattern characteristic of fluoride toxicity at the tip and along the margins, diminishing toward the midrib (Courtesy D. Tucker).

Figure 123. Perchlorate toxicity characterized by bright yellow-orange tips on sun exposed older leaves.

FURTHER READING

AMAT, S. R. 1988. Defectos y Alteraciones de los Frutos Citricos en su Comercializacion, 2nd Ed. Lit. Nicolau, Almassora, Spain.

BEATTIE, B. B., McGLASSON, W. B. and WADE, N. L. 1989. Postharvest Diseases of Horticultural Produce, Vol. 1: Temperate Fruit. CSIRO, E. Melbourne, Victoria, Australia.

DA CUNHA, G. A. P., DAVENPORT, T. L., SOULE. J. and CAMPBELL, C. W. 1978. Fruit turgor influences susceptibility of 'Tahiti' lime to stylar-end breakdown. J. Am. Soc. Hortic. Sci. 103(5):622-625.

DESZYCK, E. J. and TING, S. V. 1960. Grapefruit investigations: I. Blue albedo in grapefruit. Citrus Station Mimeo Series 61-2, Univ. of Florida, Lake Alfred.

EBLING, W. 1950. Subtropical Entomology. Lithotype Process Co., San Francisco.

FAWCETT, H. S. 1936. Citrus Diseases and Their Control, 2nd Ed. McGraw-Hill Book Co., New York.

FAWCETT, H. S., PERRY, J. C. and JOHNSTON, J. C. 1944. The stubborn disease of citrus. California Citrograph 29(6): 146-147.

FLINT, M. L. 1991. Integrated Pest Management for Citrus, 2nd Ed. Univ. of California, Davis.

GRIERSON, W., McCORNACK, A. A. and HAYWARD, F. W. 1965. Tangerine handling. Fla. Agric. Ext. Circ. 285.

JACKSON, L. K. 1991. Citrus Growing in Florida. Univ. Presses of Florida, Gainesville.

KLOTZ, L. J. 1973. Color Handbook of Citrus Diseases. Univ. of California, Riverside.

KNAPP, J. L. 1988. Florida Citrus Integrated Pest and Crop Management Handbook. Univ. of Florida, Gainesville.

KNAPP, J. L. 1995. Florida Citrus Pest Management Guide (annual). Univ. of Fla., Gainesville, FL.

KNORR, L. C. 1973. Citrus Diseases and Disorders. Univ. of Fla. Press, Gainesville, FL.

McCORNACK, A. A. and BROWN, G. E. 1991. Market diseases and blemishes of Florida citrus fruit. Fla. Dept of Citrus, Lakeland.

McCORNACK, A. A. and GRIERSON, W. 1965. Practical measures for control of stem-end rind breakdown of oranges. Fla. Agric. Ext. Circ. 286.

McPHEE, J. 1966. Oranges. Farrar, Straus and Giroux, New York.

OFFERS, J. A. 1987. Citrus Diseases and Defects Found in the Marketplace. J. A. Offers, Papendrecht, Holland.

OHR, H. D. and ECKERT, J. W. 1985. Postharvest diseases of citrus fruits in California, Univ. of Calif. Coop. Ext. Leaflet 21407.

PRATT, R. M. 1958. Florida Guide to Citrus Insects, Diseases and Nutritional Disorders in Color. Univ. of Florida, Gainesville.

REUTHER, W., BATCHELOR, L. D. and WEBBER, H. J. 1967. The Citrus Industry, Vol. I: History, World Distribution, Botany and Varieties. Univ. of Calif.

REUTHER, W., BATCHELOR, L. D. and WEBBER, H. J. 1968. The Citrus Industry, Vol. II: Anatomy, Physiology, Genetics and Reproduction. Univ. of Calif.

REUTHER, W. 1973. The Citrus Industry, Vol. III: Production Technology. Univ. of Calif.

REUTHER, W., CALAVAN, E. C. and CARMAN, G. E. 1978. The Citrus Industry, Vol. IV: Crop Protection. Univ. of Calif.

REUTHER, W., CALAVAN, E. C. and CARMAN, G. E. 1989. The Citrus Industry, Vol. II: Crop Protection, Postharvest Technology and Early History of Citrus in California. Univ. of Calif.

SAUNT, J. 1990. Citrus Varieties of the World. Sinclair International, Ltd., Norwich, England.

SNOWDON, A. L. 1990. A Color Atlas of Post-Harvest Diseases & Disorders of Fruits & Vegetables, Vol. 1: General Introduction & Fruits. Wolfe Scientific Publications, London, England.

WARDOWSKI, W. F. and BROWN, G. E. 1993. Postharvest decay control recommendations for Florida citrus fruit. Fla. Agric. Ext. Circ. 359-A.

WARDOWSKI, W. F., McCORNACK, A. A. and GRIERSON, W. 1976. Oil spotting (oleocelosis) of citrus fruit. Fla. Agric. Ext. Circ. 410.

WARDOWSKI, W. F., NAGY. S. and GRIERSON, W. 1986. Fresh Citrus Fruits. AVI Publishing Co., Westport, CT.

WARDOWSKI, W. F. and TING, S. V. 1983. Blue albedo. Packinghouse Newsletter No. 131, Univ. of Florida, Lake Alfred.

WEIR, R. G. and CRESSWELL, G. C. 1993. Plant Nutrient Disorders, Vol. 1: Temperate and Subtropical Fruit and Nut Crops. Butterworth-Heinemann, Melbourne, Australia.

WHITESIDE, J. O., GARNSEY, S. M. and L. W. TIMMER. 1988. Compendium of Citrus Diseases. Amer. Phytopath. Soc., St. Paul, MN.

INDEX

A

B

C

D

E

F

G

H

I

K

L

Q

R

S

T

U

V

W

X

Y

Z